GENOMICS
Commercial Opportunities from a Scientific Revolution

T0227993

GENOMICS
Commercial Opportunities from a Scientific Revolution

Edited by

G.K. Dixon[a], L.G. Copping[b] and D. Livingstone[c]

[a]*ZENECA Pharmaceuticals, Macclesfield, UK*
[b]*Saffron Walden, Essex, UK*
[c]*ChemQuest, Steeple Morden, UK*

Papers presented at the conference on Genomics, Commercial Opportunities from a Scientific Revolution, University of Cambridge, Cambridge, UK 30 June – 2 July 1997

Taylor & Francis
Taylor & Francis Group
LONDON AND NEW YORK

© Taylor & Francis, 1998

First published 1998

A CIP catalogue record for this book is available from the British Library.

Papers presented at the conference on Genomics, Commercial Opportunities from a Scientific Revolution, University of Cambridge, Cambridge, UK, 30 June-2 July 1997

ISBN 1 85996 106 1

Published by Taylor & Francis
2 Park Square, Milton Park, Abingdon, Oxon, OX14 4RN
270 Madison Ave, New York NY 10016
World Wide Web home page: http://www.taylor&francis.com/

Transferred to Digital Printing 2006

CONTENTS

Preface

The field of "genomics" has emerged over the last decade as one of the most exciting developments in biology this century. It is now possible to unravel the genetic code, the basic unit of life, of prokaryotic and eukaryotic organisms with the advancements in DNA sequencing technology. The first eukaryotic sequence to be completed was that of the yeast, *Saccharomyces cerevisiae*; the complete human sequence is expected to be completed by 2005. The availability of this sequence information is only the first stage in our increased understanding of the biology of the organisms concerned. Determining the functionality of genes and their expression will be necessary if the sequencing efforts are to yield improvements in medicine and agriculture. Genomics has been hailed as a scientific revolution. Is this a revolution that will deliver commercial opportunities ? How are companies making use of this information ? How quickly will genomic research have an impact in medicine and agriculture ? These are just some of the questions to be answered during the next century.

This book records the proceedings from a symposium organised by the Society of Chemical Industry and held at the Chemistry Department, University of Cambridge, 30th June - 2nd July, 1997. This was an international conference attracting over 130 delegates with the majority representing the pharmaceutical and agrochemical industries.

The conference focused on the key technologies and developments in the field of genomics with contributions from those who are leading the research internationally. It is expected that this book will become a reference text for professionals and managers in the pharmaceutical and agrochemical industries and students with an interest in this area.

Graham K Dixon, Leonard G Copping, David J Livingstone
Organising Committee

Acknowledgement

We would like to thank Zeneca Pharmaceuticals for their generous support of the conference.

SCI   Connecting Science and Industry Worldwide

http://sci.mod.org

With members in over 65 countries across the world, **SCI**, a not-for-profit association, exists to improve the exchange of information an understanding between researchers, industrialists, financiers and academics from all branches of science and all sectors of science related industry.

Learned and non-partisan, **SCI** attracts men and women from all levels of business, research and public life by providing them with access to independent and informed expertise through specialist conferences and meetings, public *Open Forum* debates, peer-reviewed journals, books, *Email Discussion Forums*, and *Chemistry & Industry* magazine.

SCI, Society of Chemical Industry
International Headquarters
14/15 Belgrave Square
LONDON SW1X 8PS, UK

Telephone: +44 (0) 171 235 3681
Fax: +44 (0) 171 823 1698
Email: secretariat@chemind.demon.co.uk
Internet: http://sci.mond.org; http://ci.mond.org

SEQUENCING PROGRAMS

SEQUENCING OF THE HUMAN GENOME

DAVID R. BENTLEY

*The Sanger Centre, Wellcome Trust Genome Campus, Hinxton,
Cambridge, CB10 1SA, UK*

ABSTRACT

The international programme to determine the 3000Mb sequence of the human genome
started in earnest in 1995. At that stage an initial framework was provided by a genetic
map, constructed using polymorphic microsatellite markers. This is being supplemented by
long range physical maps built using either radiation hybrid mapping or overlapping sets of
yeast artificial chromosome (YAC) clones. The resulting high densities of landmarks
ordered along each chromosome have made it possible to adopt a coordinated approach in
which participants of the programme select specific regions or individual chromosomes for
construction of sequence-ready maps of overlapping bacterial clones. A minimally
overlapping set is chosen for sequencing using the synthetic chain terminator method
pioneered by Sanger and co-workers (1977). The most common approach is to subclone a
100-200kb bacterial clone and to assemble the sequences of several thousand fragments in
an initial random "shotgun" phase, followed by a directed "finishing" phase to close gaps
and resolve ambiguities. An alternative approach has also been used which involves
mapping the subcloned fragments in place of the random phase before sequencing.

The agreed aim is to produce finished genomic sequence at an accuracy of greater than
99.99% and to leave no gaps. To assist coordination, assessment of progress, and to
provide early access to valuable information, unfinished assembled sequence is made
available as a pre-release in the public domain. All sequence is then finished, annotated
and submitted to the public databases. The target date for completion is 2005.

The nucleotide sequence of the human genome, in concert with that of other organisms, will
provide us with the foundation for the next era of research in biology and medicine. It will
draw together all the existing genomic information and provide us with a complete
description of all the genes and their homologous relationships. This information will assist
our understanding of the genetic determination of biochemical and developmental
processes, and our ability to combat human disease.

INTRODUCTION

The description of the structure of DNA by Crick and Watson (1953) defined how
biological information is stored digitally in the linear order of the four bases in the double
helix. As a result it became possible to consider recording and decoding the entire
complement of genetic information that determines any living organism. Knowledge of the

DNA structure also provided the basis for elucidating the mechanism of DNA synthesis. This in turn led to the development of the *in vitro* synthetic DNA sequencing method by Fred Sanger and co-workers (1977) that is now used in all large-scale genomic sequencing.

Knowledge of the entire sequence of a genome provides access to the blueprint that determines structure, function and development of the organism. In general, organisms of greater complexity have larger genomes which encode more genes. Thus an example of a prokaryotic unicellular organism is the enteric bacterium *Escherichia coli*. *E.coli* has a genome of 5 megabases (5Mb) encoding 4,300 genes. The genome of a simple unicellular eukaryote, *Saccharomyces cerevisiae* (bakers yeast), is three times larger (15Mb) and encodes 6,000 open reading frames. The sequence of both these genomes is now complete (Blattner *et al.*, 1997; Goffeau *et al.*, 1997). The genome of the nematode worm *Caenorhabditis elegans* (100Mb) encodes at least 13,000 genes; the sequence is due for completion next year (reviewed by Waterston and Sulston, 1995). *C. elegans* is a multicellular eukaryote with approximately 1,000 cells (more precisely the surviving somatic nuclear count is 1031 in the adult male) and displays a significant degree of cellular differentiation. The genome, therefore, contains not only genes which determine basic eukaryotic cell functions (many of which are similar in yeast), but also genes which determine many other functions that are necessary for cell specialisation and multicellularity. The human genome comprises approximately 3,000Mb. Estimates of the total number of human genes range from 80-150,000. While the human gene complement determines a complex, highly differentiated multicellular organisation, it also includes genes that encode many of the same basic functions found in yeast and the nematode worm. Thus the sequence of the human genome should not be viewed in isolation, but rather in the context of our knowledge of other genomes and the cellular and biochemical functions encoded by genes which may have homologues in many species.

GENOME MAPPING

The mapping and sequencing of other genomes has played an important part in the development of methods and strategies that have since been applied to the human genome. The physical maps of the yeast and nematode genomes were constructed by identifying strings of overlapping clones. Initially, restriction digestion was used to generate unique fingerprints of bacterial clones picked at random. Overlaps between clones were then established by comparison of the fingerprint patterns, and the clones were assembled into contiguous sections (contigs) (Coulson *et al.*, 1977; Olson *et al.*, 1977). Gaps between contigs were bridged using YAC clones in the case of the *C. elegans* map; Coulson *et al.*, 1988), or by extending existing contigs by isolating further bacterial clones in a stepwise manner where possible (chromosome walking). The maps of both genomes reached an advanced state prior to the start of systematic large-scale sequencing.

In the early stages of mapping the human genome, the emphasis was in ordering unique landmarks along the chromosomes. Olson *et al.* (1989) observed that any unique position in a genome could be defined, or tagged, purely by knowledge of its DNA sequence, hence sequence tagged site (STS). Further, the development of the polymerase chain reaction (PCR) (Saiki *et al.*, 1985, 1988) provided a convenient assay for any STS, as the sequence could be used to design a pair of primers which would amplify a product of defined size only if the target DNA sequence was present in the sample being tested (STS content mapping; reviewed by Green and Green, 1991). The information required to define both the STS itself and a specific assay could be communicated electronically. The use of STSs to define genomic landmarks expanded rapidly, and soon exceeded the use of hybridisation probes in mapping. STSs could define any type of landmark, such as genes or individual exons, genetic markers or anonymous sections of genomic DNA. This, coupled with the convenience of scaling up the PCR assay, formed the basis for carrying out the large-scale mapping that was required to tackle the human genome.

An important milestone in mapping the human genome was the construction of the first genome-wide genetic linkage map by Weissenbach *et al.* (1992). The map was constructed by using polymorphic microsatellite sequences (direct tandem repeats of the dinucleotide CA; Weber and May, 1989) as genetic markers. Large-scale typing of a panel of DNA samples from eight extended pedigrees was carried out for each marker using PCR assays. The final map contained 5,264 markers, of which 2,032 were ordered at odds of >1000:1 (Dib *et al.*, 1996). This provided the first framework map of landmarks across the human genome.

Initial attempts to construct genome-wide physical maps were made using YAC clones containing inserts of average size 1Mb. These attempts (Cohen *et al.*, 1993; Chumakov *et al.*, 1995) were only partly successful as the physical distances between the genetic markers that were available at the time was, in general, too great to be spanned by single YAC clones. Strategies to link clones together in overlapping sets in order to bridge the distances met with several difficulties. Notably, a high proportion of the clones in the large insert human YAC libraries was found to be chimaeric, i.e. they contained an insert with DNA derived from two or more discontinuous regions of the genome. This made it impossible to use the YAC clones to cover extensive tracts of the genome in the absence of additional information to verify the map.

Intensive studies of specific chromosomes met with more success, as information from a variety of additional sources, including fluorescent *in situ* hybridisation to chromosome spreads, pulsed-field gel electrophoresis, analysis of somatic cell hybrid lines, and other genetic maps was used in the verification process. These efforts resulted in the construction of YAC-based maps for the euchromatic portion of the Y chromosome (Foote *et al.*, 1992), chromosome 21 (Chumakov *et al.*, 1992), followed by chromosomes 22 (Collins *et al.*, 1995), 16 (Doggett *et al.*, 1995), 3 (Gemmill *et al.*, 1995), 19 (Ashworth *et al.*, 1995), 12 (Krauter *et al.*, 1995) and 7 (Bouffard *et al.*, 1997). These maps provided both overlapping clones, and sets of ordered landmarks at higher densities (ranging from 1/250 to 1/70 kb) than were available from the genetic maps. In contrast to the genetic maps, construction of YAC-based maps also allowed the integration of different classes of landmark, including genes and anonymous genomic fragments.

The development of radiation hybrid (RH) mapping (Goss and Harris, 1975), applied initially to single chromosomes and then to the whole genome (Cox *et al.,* 1990; Walter *et al.,* 1994), provided an alternative means to develop high density framework maps. This method was used initially to complement and extend genome-wide YAC maps (Hudson *et al.,* 1995) and more recently to construct high-density framework maps independently of YACs. Like the earlier genetic linkage map, RH mapping employed large-scale PCR typing of a panel of DNA samples (in this case RH cell lines), but higher resolution maps were obtained, and any amplifiable landmark (including the microsatellite markers from the genetic map) could be integrated into the RH map. This approach has been used in the construction of the human gene map of expressed sequence tags (ESTs); the first compilation of the gene map included 16,000 genes (Schuler *et al.,* 1996), and the latest map contains approximately 30,000 genes (P. Deloukas, personal communication). RH mapping is also being used to develop landmark maps at a density of up to 1/70kb on selected chromosomes (1, 6 and 20). This forms the basis for our current strategy to make sequence-ready maps, which is summarised in fig. 1.

Two new bacterial cloning systems, the P1- and bacterial artificial chromosomes (PACs and BACs, respectively) have been developed which enable cloning of DNA fragments with an average size of up to 250 kb (Shizuya *et al.,* 1992 ; Ioannou *et al.,* 1994). The cloned DNA insert is maintained as a single copy in the host cell, in contrast to the multiple copy cosmid and fosmid systems, and the PACs and BACs are more stable and easier to work with than the earlier bacterial or yeast genomic clones. PACs and BACs, therefore, now constitute the preferred clones for large-scale sequencing. Large libraries of these bacterial clones are screened using all the available landmarks in a genomic region. The screening is done by hybridisation of gridded high-density arrays on nylon filters, or by PCR screening of DNA representing pools of clones. The positive clones are isolated and assembled into contigs on the basis of their landmark content and restriction fingerprints. The contigs are then extended and joined by walking. One of several protocols is used to isolate a DNA fragment or sequence from the ends of the appropriate clones. A new assay is then developed from each end to identify a join with a neighbouring contig, or to screen for new clones that will extend the contig.

Once the clone map is confirmed, a minimally overlapping subset of clones is selected for sequencing. Prior to subcloning and sequencing, our procedure in-house is to extract the DNA from each PAC or BAC (or cosmid or fosmid) clone and to verify it by refingerprinting and fluorescent *in situ* analysis .

GENOMIC SEQUENCING

The first step in determining the complete nucleotide sequence of a 100-200kb clone is to fragment the purified DNA randomly by shearing or sonication. DNA fragments are incubated with mung bean nuclease which renders the termini blunt-ended. After size separation on agarose, 1.4-2.2kb fragments are subcloned into a plasmid or M13

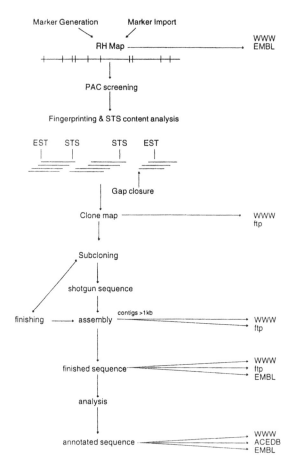

FIGURE 1. Strategy for mapping and sequencing the human genome. Markers are imported from public database or generated *de novo*, for example from sequences of subclones of flow-sorted chromosomes, and integrated to form a single framework map by analysis of radiation hybrids (as illustrated) or overlapping YACs. The markers are then used to isolate bacterial clones which are ordered by fingerprinting and STS content analysis, and gaps between contigs are closed by chromosome walking to complete the bacterial clone map. PAC, BAC cosmid or fosmid clones are selected for sequencing, and a preparation of insert DNA is sheared and subcloned into bacteriophage M13 or a plasmid vector to provide templates for shotgun sequencing. Following assembly of the random sequences, several rounds of finishing are carried out to close all gaps, resolve ambiguities, and check all the data. The finished sequence is analysed and annotated in an implementation of the database ACEDB prior to submission. The release of data on the internet (http//:www.sanger.ac.uk) at various stages of the process is indicated. In addition, RH data is submitted to rhdb at the EBI, the clone maps, and sequence data are available via ftp, and both unfinished sequence, and the finished, annotated product, are available at EBI, NCBI and DDBJ:
(see http//:www.ebi.ac.uk, http//:ncbi.nlm.nih.gov and http://www.ddbj.nig.ac.jp/).

bacteriophage vector. DNA of each subclone is extracted for use as a template in the synthetic chain terminator sequencing reaction.

The sequencing itself proceeds in two phases: an initial random shotgun phase, followed by a phase of directed finishing. In the shotgun phase, the end sequence of approximately 3,000 templates (for a 100kb PAC) is determined in individual sequencing reactions; plasmid (pUC) templates are sequenced from both ends routinely. During the reaction, a specific fluorescent tag is incorporated into the nascent DNA chain in one of two ways. In dye-primer sequencing, a universal sequencing primer with an attached fluorescent dye of a specific colour is used to prime the *in vitro* DNA synthesis. Four reactions are set up in parallel for each template, each with a different chain terminator dideoxynucleoside triphosphate (ddNTP) and corresponding dye-labelled universal primer. The products of the four reactions are pooled, and then electrophoresed and analysed automatically using an automated fluorescent gel reader. In dye-terminator sequencing, each of the four dyes is attached directly to the corresponding ddNTP. The sequencing of each template is done in a single tube and the products are loaded on to the gel reader. In general, dye-primer sequencing gives longer read lengths, while the dye-terminator sequencing has the ability to resolve some compressions (which are caused by abnormal migration of different DNA fragments in the same place in the gel - see example in fig. 2) better than the dye-primer protocol.

The sequences obtained in the initial shotgun phase are automatically compared and aligned. Matches detected between individual sequences determine their overlapping relationship, and the sequences are assembled into contigs (see example in Fig. 2). The shotgun sequence usually assembles into 5-30 contigs representing 95% of the 200kb PAC or BAC clone. At this stage, the data is made available via the internet (see fig. 1 for details).

In the directed finishing phase, the product of the initial assembly is subjected to a number of steps with the aim of closing the remaining gaps between contigs, resolving ambiguities, and checking the entire sequence. Gap closure can be achieved by obtaining a longer read of the same template to cross the gap, by sequencing the same template from the other end, or by extending the sequence further along a template using a specific oligonucleotide as a new primer. Ambiguities are resolved either by inspecting and editing the existing assembled data manually, or by performing additional sequencing reactions to obtain more conclusive raw data. Additional reactions are also carried out to ensure that every base has been determined independently more than once. Wherever possible, every base is determined from two independent templates or with two reaction chemistries. Some of the most difficult problems to solve are gaps in GC-rich regions, often where the sequence allows the formation of a small hairpin loop that interferes with chain extension by the DNA polymerase. In these cases, multiple additional subclones with short inserts are prepared and sequenced. Sequences of subclones which do not contain all of the hairpin provide the necessary data for finishing. A completed sequence has an accuracy equal to or greater than 99.99%, and has no gaps.

(a)

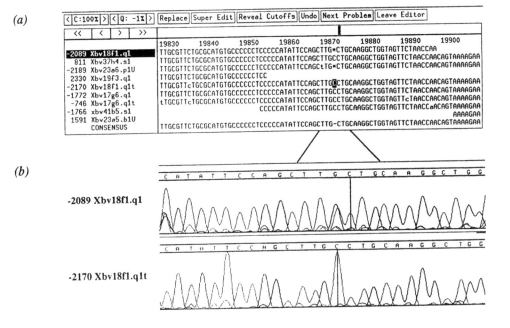

(b)

FIGURE 2. Assembled unfinished sequence in an XGAP database. (a) An example of unfinished sequence that has been assembled using the program PHRAP and subsequently displayed in the editor program XGAP. Each line represents the sequence data obtained from one template. Lower case letters in individual lines indicate positions where the the sequence has been edited either automatically (highlighted in buff) or manually (highlighted in red) on the basis of the data in the other reads; these are then easily spotted during the checking process. The asterisks occur in the same position in two reads where there is an ambiguity that requires further analysis. (b) Gel traces of two of the reads from (a), using different reaction chemistries and the same template, highlighting an ambiguity. In the upper trace, which is a dye primer reaction, a compression of the bands in the gel has resulted in co-migration of a peak representing a C residue with the peak representing the adjacent G. As the G peak is larger, a G nucleotide has been called by the base-calling algorithm. From visual inspection of the trace, however, both peaks are clearly visible. In the lower trace, the use of the dye terminator chemistry (as indicated by the suffix 't' in the template name) results in absence of any compression in the equivalent region of sequence. The peaks of the G and the following C are well separated, and both bases are called by the program.

ANALYSIS

Computational analysis is the first step in interpreting the information contained within the genomic sequence. It allows detection of many features of biological importance, and also serves the important function of drawing all the information about a particular region that has previously been obtained together on a single metric. Genes can be identified by a number of methods, and the results of the analysis will contain all known STSs, including polymorphic markers.

Following masking of the known repeat elements, the finished sequence is used to search for matches with DNA sequences in all the publicly available databases, including all EST and cDNA information, and available sequence data from other species. Matches with known cDNA sequences and ESTs indicate the presence of genes; homology (but not identity) to other genes or proteins may also indicate the presence of genes. Genes may also be predicted *de novo* using programs such as Genescan (Burge and Karlin, 1997), FGeneh, Grail, Hexon and others (evaluated in Burset and Guigo, 1996). Further useful evidence for the existence of genes in genomic sequence is the presence of potential CpG islands (CGIs). CGIs are regions (usually at least 1kb in length) of high GC content with a CpG:GpC ratio of approximately 1:1, in contrast to the rest of the genome which has a CpG:GpC ration of 1:4. 60% of all genes in the human genome are associated with CGIs, which occur nearly always at the 5' end of the gene and are unmethylated (Bird, 1986).

Following analysis, the finished sequence of each bacterial clone is annotated prior to submission in the public databases at the European Bioinformatics Institute (the European Molecular Biology Laboratory (EMBL) database), the National Centre for Biotechnology Information (NCBI Genbank) and DNA Data Bank of Japan (DDBJ). All submitted data is stored at all three sites, ensuring world wide availability of the sequence.

THE INTERNATIONAL PROGRAMME

The international programme to determine the sequence of the human genome began in earnest in 1995. The tenets of the proposal were that knowledge of the complete sequence would draw all genetic information together, providing a basis for a complete description of all the genes and other features, and also the necessary resources (chiefly the sequence itself) to underpin research in biology and medicine in the next century. An agreement between participating laboratories has been made which states that "all human genomic sequence information, generated by centres funded for large-scale human sequencing, should be freely available and in the public domain in order to encourage research and development and to maximise its benefit to society" (Report of the First International Strategy Meeting on Human Genomic Sequencing, 1995).

As of September 1997, approximately 60Mb (2%) of the human genome sequence has been completed, the largest contributions coming from the Sanger Centre, Cambridge, UK (25Mb) and the Genome Sequencing Centre, St Louis, USA (16Mb). Between one-third and one-half of the human genome is actively being worked on at the level of constructing framework maps, bacterial clone maps, or sequencing. The stated goal is to complete the task early in the next century, an exciting challenge for the genome sequencing community. To understand the information stored within it will take a little longer.

ACKNOWLEDGEMENTS

The work reviewed here involves the contribution of many laboratories to the international programme. It takes particular account of the work of all the staff at the Sanger Centre, Cambridge, UK, (funded by the Wellcome Trust and the UK Medical Research Council) and the Genome Sequencing Centre, St Louis, USA, (funded by the National Institutes of Health) on whose behalf this paper is written. I am also grateful to Mark Vaudin, Richard Durbin and John Sulston for critical reading of the manuscript, and Jane Wilkinson, Gill Hannam and Linda Mansfield for assisting with the figures and text.

REFERENCES

Ashworth L.K., Batzer M.A., Brandriff B., Branscomb E., de Jong P., Garcia E., Garnes J.A., Gordon L.A., Lamerdin J.E., Lennon G., Mohrenweiser H., Olsen A.S., Slezak T. and Carrano A.V. (1995). An integrated metric physical map of human chromosome 19. *Nature Genetics* **11**:422-427.

Bird, A. P. (1986). CpG-rich islands and the function of DNA methylation. *Nature* **321**:209-13.

Blattner F.R., Plunkett III G., Bloch C.A., Perna N.T., Burland V., *et al.* (1997) The complete genome sequence of *Escherichia coli* K-12. *Science* **277**:1453-1474.

Bouffard G.G., Idol J.R., Braden V.V, Iyer L.M., Cunningham A.F., Weintraub L.A., Touchman J.W., Mohr-Tidwell R.M., Peluso D.C., Fulton R.S., Ueltzen M.S., Weissenbach J., Magness C.L., and Green E.D. (1997). A physical map of human chromosome 7: an integrated YAC contig map with average STS spacing of 79kb. *Genome Research* **7**:673-692.

Burge C. and Karlin S. (1997) Prediction of complete gene structures in human genomic DNA. *Journal of Molecular Biology* **268**:78-94.

Burset M. and Guigo R. (1996). Evaluation of gene structure prediction programs. *Genomics* **34**:353-367.

Chumakov I.M., Rigault P., Le Gall I., Bellanné-Chantelot C., Billault A., Guillou S., Soularue P., Guasconi G., Poullier E., Gros I. *et al.* (1995). A YAC contig map of the human genome. *Nature* **377**:175-298.

Chumakov I., Rigault P., Guillou S., Ougen P., Billaut A., Guasconi G., Gervy P., LeGall I., Soularue P., Grinas L. and et al. (1992). Continuum of overlapping clones spanning the entire human chromosome 21q. *Nature* **359**:380-7.

Cohen D., Chumakov I., and Weissenbach J. (1993). A first-generation physical map of the human genome. *Nature* **366**:698-701.

Collins J.E., Cole C.G., Smink L.J., Garrett C.L., Leversha M.A., Soderlund C.A., Maslen G.L., Everett L.A., Rice K.M., Coffey A.J.*et al.* (1995). A high-density YAC contig map of human chromosome 22. *Nature* **377 suppl.**:367-379.

Coulson A., Sulston J., Brenner S. and Karn J. (1986). Towards a physical map of the genome of the nematode *C. elegans*. *Proceedings of the National Academy of Sciences U.S.A* . **83**:7821-7825.

Coulson A., Waterston R., Kiff J., Sulston J., and Kohara Y. (1988). Genome linking with yeast artificial chromosomes. *Nature* **335**:184-6.

Cox D.R., Burmeister M., Price E.R., Kim S., and Myers R.M. (1990). Radiation hybrid mapping: a somatic cell genetic method for constructing high-resolution maps of mammalian chromosomes. *Science* **250**:245-50.

Dib C., Faure S., Fizames C., Samson D., Drouot N., Vignal A., Millasseau P., Marc S., Hazan J., Seboun E., *et al.* (1996). A comprehensive genetic map of the human genome based on 5,264 microsatellites. *Nature* **380**:152-154.

Doggett N.A., Goodwin L.A., Tesmer J.G., Meincke L.J., Bruce D.C., Clark L.M., Alther M.R., Ford A.A., Chi H-C, Marrone B.L. *et al.* (1995). An integrated physical map of human chromosome 16. *Nature* **377 suppl.**: 335-364.

Foote S., Vollrath D., Hilton A. and Page D.C. (1992). The human Y chromosome: overlapping DNA clones spanning the euchromatic region. *Science* **258**:60-6.

Gemmill R.M., Chumakov I., Scott P., Waggoner B., Rigault P., Cypser J., Chen Q., Weissenbach J., Gardiner K., Wang H., *et al.* (1995). A second generation YAC contig map of human chromosome 3. *Nature* **377 suppl.**: 299-333.

Goffeau A., Aert R., Agostini-Carbone M.L., Ahmed A., Aigle M., *et al.*(1997) The yeast genome directory. *Nature* **387 suppl.**:1-105.

Goss S. J. and H. Harris. (1975). New method for mapping genes in human chromosomes. *Nature* **255**:680-4.

Green E.D. and Green P. (1991). Sequence-tagged site content mapping of human chromosomes: theoretical considerations and early experiences. *PCR Methods and Applications* **1**:77-90.

Hudson T.J., Stein L.D., Gerety S.S, Ma J., Castle A.B., Silva J., Slonim D.K., Baptista R., Kruglyak L., Zu S-H, X. Hu *et al..* (1995). A sequence-tagged site based map of the human genome. *Science* **270**:1945-1954.

Ioannou P.A., Amemiya C.T., Garnes J., Kroisel P.M., Shizuya H., Chen C., Batzer M.A. and de Jong P.J. (1994). A new bacteriophage P1-derived vector for the propagation of large human DNA fragments. *Nature Genetics* **6**:84-9.

Krauter K., Montgomery K., Yoon S-J., LeBlanc-Straceski J., Renault B., Marondel I., Herdman V., Cupelli L., Banks A., Lieman J. *et al.* (1995). A second generation yeast artificial chromosome contig map of human chromosome 12. *Nature* **377 suppl.**: 321-333.

Olson M.V., Dutchik J.E., Graham M.Y., Brodeur G.M., Helms C., Frank M., MacCollin M., Scheinman R. and Frank T. (1986). Random-clone strategy for genomic restriction mapping in yeast. *Proceeding of the National Academy of Sciences U.S.A.* **83**:7826-30.

Olson M., Hood L., Cantor C. and Botstein D. (1989). A common language for physical mapping of the human genome. *Science* **245**:1434-5.

Saiki R.K., Gelfand D.H., Stoffel S., Scharf S.J., Higuchi R., Horn G.T., Mullis K.B. and Erlich H.A. (1988). Primer-directed enzymatic amplification of DNA with a thermostable DNA polymerase. *Science* **239**:487-91.

Saiki R.K., Scharf S., Faloona F., Mullis K.B., Horn G.T., Erlich H.A. and Arnheim N. (1985). Enzymatic amplification of beta-globin genomic sequences and restriction site analysis for diagnosis of sickle cell anemia. *Science* **230**:1350-4.

Sanger F., Nicklen S. and Coulson A.R. (1977). DNA sequencing with chain-terminating inhibitors. *Proceeding of the National Academy of Sciences U.S.A.* **74**: 5463-5467.

Shizuya H., Birren B., Kim U.J., Mancino V., Slepak T., Tachiiri Y. and Simon M. (1992). Cloning and stable maintenance of 300-kilobase-pair fragments of human DNA in Escherichia coli using an F-factor-based vector. *Proceeding of the National Academy of Sciences U.S.A.* **89**:8794-7.

Walter M. A., Spillett D.J., Thomas P., Weissenbach J. and Goodfellow P.N. (1994). A method for constructing radiation hybrid maps of whole genomes. *Nature Genetics* **7**:22-8.

Waterston R. and Sulston J. (1995) The genome of *Caenorhabditis elegans*. *Proceeding of theNational Academy of Sciences U.S.A.* **92:**10836-10840.

Watson J.D and Crick F.H.C. (1953) Molecular structure of nucleic acids. A structure for deoxyribose nucleic acid. *Nature* **171:**737-738.

Weber J. L. and May P.E. (1989). Abundant class of human DNA polymorphisms which can be typed using the polymerase chain reaction. *American Journal of Human Genetics* **44:**388-96.

Weissenbach J., Gyapay G., Dib C., Vignal A., Morissette J., Millasseau P., Vaysseix G. and Lathrop M. (1992). A second-generation linkage map of the human genome. *Nature* **359:**794-801.

GENOME PROJECTS: FROM MICROBES TO MAN, FROM SEQUENCE TO FUNCTION

JOHN QUACKENBUSH, MARK D. ADAMS, CLAIRE M. FRASER AND J. CRAIG VENTER

The Institute for Genomic Research, 9712 Medical Center Drive, Rockville, MD 20850, USA

ABSTRACT

The determination of the complete nucleotide sequence of the Human Genome is a goal that has captivated both scientists and the lay public for nearly a decade. We are rapidly progressing toward the determination of that sequence and the world-wide effort to sequence the human genome has set 2005 as the target date for the completion of that goal. The promise of the Genome Project has been that, as the sequence unfolds, we will be able to obtain a catalog of all human genes, as well as to begin investigation of the large-scale structure and organization of the genome. Comparisons within and between species promise to provide substantial information about the function of many genes. Knowledge of the sequence of the human genome will greatly enhance our ability to detect sequence variation among individuals. This, in turn, will be invaluable for identifying genes underlying human diseases, as sequence variation provides the necessary link between genes and phenotypes. Knowledge of the gene sequences alone, however, provides only part of the puzzle. The challenge for the future is to identify those genes that are responsible for the development of a specific cellular or organismal phenotype. Scientists at The Institute for Genomic Research have been leaders in genomic research, including both Expressed Sequence Tag and genomic sequencing, and are building a program that we hope will define the manner in which future researchers will address questions of gene function using genomic techniques.

INTRODUCTION

The Human Genome Project is a major international effort having the sequencing of the complete human genome and the identification of the entire complement of human genes as its primary mission. However, early on there was a realization that the sequencing of the human genome in isolation was not sufficient; the identification of expressed human sequences and the sequencing of model organisms were essential for the project's success (Collins and Galas, 1992). This world-wide effort represents the first coordinated effort to access and understand the complex information stored within the Genome.

The Institute for Genomic Research (TIGR) is a non-profit research institute with interests in structural, functional, and comparative analysis of genomes and gene products in viruses, eubacteria, pathogenic bacteria, archaea, and eukaryotes, both plant and animal, including humans. TIGR has long realized the value of genomic approaches and has made significant contributions to the Genome Project. Researchers at TIGR developed Expressed Sequence Tag (EST) sequencing for the discovery and mapping of human genes and were first to apply high-throughput, automated sequencing techniques to the collection of EST data (Adams *et al.*, 1991; Adams *et al.*, 1995). TIGR has made significant contributions to genomic sequencing, including the first fully sequenced genome of a free-living organism, *Haemophilus Influenzae* Rd (Fleischmann *et al.*, 1995), and the first completely sequenced member of the Archaea, *Methanococcus jannaschii* (Bult *et al.*, 1996). TIGR has since completed the sequencing of four additional prokaryotes and is actively sequencing twelve others. TIGR is also making significant contributions to the sequencing of eukaryotes, including *Plasmodium falciparum*, *Arabidopsis thaliana*, and humans.

Complete genomic sequence is only the first step in developing an understanding of fundamental processes underlying life. The challenge for the future is to define the functions of the proteins encoded within the genome as well as the functional networks that they comprise. TIGR is taking an active role in defining the Next Genome Project, moving us from sequence to function, by building a program in whole-genome functional analysis.

EXPRESSED SEQUENCE TAGS: A FIRST GLIMPSE AT FUNCTION

The sequencing of Expressed Sequence Tags (ESTs; Adams *et al.*, 1991) has proven to be an invaluable tool for gene discovery and gene mapping in humans (Affara *et al.*, 1994; Schule *et al.*, 1996) and other species (McCombie *et al.*, 1992; Waterston *et al.*, 1992) and large-scale EST sequencing projects have added an invaluable source of data to the public domain. ESTs are partial, single-pass sequences from either end of a cDNA clone. The EST strategy was developed by TIGR researchers and the application of high-throughput, automated DNA sequencing techniques allowed rapid identification of expressed genes by sequence analysis. TIGR has contributed more than 170,000 sequences to the public EST collection. These represent more than 52 million nucleotides of human DNA sequence, generated from 300 cDNA libraries constructed from 37 distinct organs and tissues. The EST approach has since been widely adopted; at present 71% of all GenBank entries and 40% of the individual bases in the database are derived from more than 1,000,000 EST sequences (Schuler, 1997).

Messenger RNA species are present in different concentrations in cells, and these differences are reflected in the composition of non-normalized cDNA libraries. Random sampling strategies result in abundant mRNAs being represented by many ESTs. When combined with information regarding the source of the library, this allows both gene expression levels and expression patterns to be estimated. The goal of TIGR's EST sequencing project was to build a catalog of human genes, the tissues in which they are expressed, and their expression levels. Comparative EST analysis such as this requires a cDNA library to be representative of the initial poly(A)$^+$ RNA population; great care was taken to assure that this was the case for the TIGR libraries. Using tissue samples obtained from the National Disease Research Interchange (Philadelphia, PA, USA) and The

International Institute for the Advancement of Medicine (Philadelphia, PA, USA), cDNA libraries were constructed from poly(A)$^+$ RNA. The quality of each library was evaluated and those containing both a high fraction of recombinants and an average insert size of greater than 1000 bp were used for final analysis. The clones representing these libraries were arrayed in 96-well microtiter plates and double-stranded templates were prepared for automated sequencing.

DNA sequencing reactions were performed using either the Applied Biosystems (ABI) Catalyst Lab station or Perkin-Elmer 9600 Thermocyclers with ABI Dye Primer reagents using M13 forward (-21M13) and M13 reverse (M13RP1) primers. These reactions were loaded on ABI 373A sequencers, data were collected, and transferred to the TIGR Expressed Sequence Tag Database (ESTDB). These sequences were subjected to rigorous quality control screening to eliminate poor quality sequences from further analysis and to remove vector and adaptor, poly(A/T) sequence, mitochondrial and ribosomal RNA, as well as low quality sequences from the end of the run. Only sequences with read lengths greater than 100 bp and containing fewer than 3% Ns were retained; the average trimmed length of these sequences was 292 bp.

Isolated EST sequences, while extremely valuable mapping reagents, often do not contain sufficient information to identify the encoded gene. cDNA libraries are generally constructed by oligo(dT) priming and consequently the 3' ends of genes tend to be overrepresented. These limitations, however, can be largely overcome if overlapping ESTs representing the same gene can be assembled to produce longer, contiguous blocks of sequence. To make the best possible estimate of the total number of genes represented by the growing body of EST sequence data generated world-wide, data from other EST sequencing projects that had been submitted to the dbEST Division of GenBank were treated to the same rigorous quality control standards as TIGR ESTs and added to the TIGR data set and treated as elements of a shotgun sequencing project to assemble Tentative Human Consensus (THC) sequences (Adams *et al.*, 1995). Complete gene sequence data, where available, were also used to supplement the EST collection and to seed assemblies. First, all *Homo sapiens* sequences in the PRIMATE division of GenBank were extracted. Non-coding sequences were discarded and cDNAs and coding sequences from genomic entries were saved. Redundant entries for the same gene were removed, but the link to the GenBank accession number was maintained. These sequences, representing a non-redundant set of Human Transcripts (HTs), as well as any related data, were stored in TIGR's Expressed Gene Anatomy Database (EGAD) and included in the assembly process.

For assembly, all processed EST sequences and non-redundant HT sequences were combined. The TIGR Assembler (Sutton *et al.*, 1995) was used to assemble these sequences. TIGR Assembler rapidly assesses sequence-level similarity by tabulating 10 bp oligonucleotide subsequences for each EST and places ESTs of similar oligonucleotide content into a search list of possible overlapping sequences. A seed sequence initiates assembly which is then extended by adding the best matching fragment from the search list to the consensus sequence. An EST sequence is added to the consensus sequence only if it meets stringent overlap criteria as determined by performing a modified Smith-Waterman alignment with the consensus sequence. These criteria include a minimum 95% identity in an overlap region of at least 40 bp with a maximum unmatched overhang of 20 bp. The result of this process is the collapse of the nearly 700,000 EST sequences generated world-wide into approximately 65,000 distinct THC sequences representing a first approximation of coding potential within the human genome.

The THC sequences along with information about their sources, expression patterns, and the clones containing the underlying sequence have been organized into the TIGR Human Gene Index (HGI), available *via* the world wide web at <http://www.tigr.org/tbd/hgi/hgi.html>. An example of a THC is shown in Figure 1.

MICROBIAL GENOME SEQUENCING: FINDING FUNCTION THROUGH SEQUENCE

The success of the THC assembly process was dependent on the development of software capable of handling large quantities of sequence data and using that data to rapidly produce high quality sequence assemblies. The development of that software, the TIGR Assembler, led to the speculation that the shotgun sequencing and assembly of an entire microbial genome might be possible. TIGR researchers applied this whole-genome shotgun sequencing approach to *Haemophilus influenzae* Rd, leading to the completion and publication of the first complete genome sequence of a free-living organism (Fleischmann *et al.*, 1995).

One small-insert and one medium-insert plasmid library were generated by random mechanical shearing of genomic DNA. From these, 24,304 DNA sequences representing 11,631,485 base pairs were collected and assembled using TIGR Assembler, resulting in 140 contigs. To order and orient contigs, a large-insert (l) library was generated and end sequences from the l-clones were assembled with those from the plasmids, allowing the l-clones to serve as a genome scaffold. Having sequence data from both ends of the double-stranded plasmid templates proved to be a great advantage since they made it possible to immediately link 98 of these contigs; these "sequence gaps" were closed by primer walking on clones linking the contigs. The remaining 42 "physical gaps" were regions not represented in the original clone set. The contigs were ordered by combinatoric PCR and Southern analysis and were closed by sequencing PCR products. Additional sequences were also collected to resolve ambiguities in the final assembled sequence. In total, 26,708 sequences were assembled to span the 1,830,137 base pair genome.

H. influenzae, a gram-negative organism whose only natural host is human, is a commensal resident of the upper respiratory tract and causes otitis media and respiratory tract infections, mostly in children. Analysis of the *H. influenzae* genomic sequence identified 1,743 Open Reading Frames (ORFs) representing likely genes encoded within the genome. Of these, predicted gene products from 1,007 could be assigned a putative function based on similarity to previously characterized proteins in other organisms. The remaining 736 ORFs (42.2%) encode "hypothetical proteins" with no known function, although nearly half

```
HCD Report Results: THC157407

EST# s are linked to HCD EST reports. HT# s are linked to EGAD HT reports.
GB# s are linked to GenBank accessions. ATCC#s are linked to order forms for
requesting clones.
-----------------------------------------------------------------------

>THC157407 THC32665 THC35347 THC62082 THC87770 THC131731 THC134885
nTTTTwTTTTcAAGTTTATAGTCTAAATTTTATTATCTCCAAGTCACAATGCTGATCACAAATGGGCACCCTTTAAAACAGTAACAAA
AAACACCACCACACATGGAAAAATCCTTGCAACTAAACACAGTGGACCAACAGAGACAACTCTCACAGTGTCTTAAGGTCTGGGAATC
TGGGCATGCTGCCaCAGGCTTGAGGAGACATCTTCAGGTTTAAGGCAAAGGGAACAGCCTACAAAAGGCACAACCACCAGCTACCCCT
AGAAGAATCTCTTAGTTATTTCCTCCTTGGGGGTTACAGATTAAGTGCCTCTCCCCACTCTCCATCCCACACCTGTGACTCAGAGTGA
TTAGGCCAGCTGCTAGATGGaAGGAATAAAAACAGTGACATTACCGGGGAGAGACACAGCCACCATCTTTGCCCTCAGGTTCTGTAGA
AGGACAGGGACAGTGGCCAGGTTACCCCTGGCAGACGTATGTACTGCAGTAATAGGAATCTCTTCCaCAGAGGCAGCAGAGAAGTGGT
TTAGTGCCaTGGATAGGGAGGAAAGATAGGAGCCCTTGCCCAGAAGGTGACTGGCTTCCTTAGGCCTTAAGCCCAGGAGTTTTTATAT
TCTGTCTGCAAGGACAAAAATAGAATTCGGGGAAAATAAGGTAGTAACATCTAAAACACTTGTAGCAGGAAAGACGTGGAGAAGAGCA
ATTGCAAAGGACAGGGTAGACTGCTTGCTGGATAATATTTGCCTTTAAAgAGATGCATTGGTCCACAGCAACTGGAAAAGGGGTGTAG
CAAGAAGAATGGAAAGAAgAGAAGCAAGGCCCTCTAATTCCACTTACCTCAAAAGACTGAGCCCTGAGGACTATGTGAATACACACCC
CTAAAGGCAAAGGCTCTCACTCCACCATCCCTTTCTCACAAAAGCATCTGGTGTGCCATCTCTCCCCACAGTAGGGTATTGGCCCTCA
GCAGAGAACAGAAGCCCA

Putative ID:

  1=============================THC157407===============================986
  -----------------1--------------->  <---------------27---------------
  <-----------------2----------------  ---  <-------------29---------------
  ----------3---------->  <--------23--------  <-------------30--------------
  <---------------------4----------------------->
  -----------------5----------------->  <-------31-------->
  <--------6-----------  <---------------24------------------
  ---------------7------------------->
  ----------------8------------------->
  -----------9----------------->  <---------------25------------------
  ----------------10--------------->
  ---------------11------------------->
  ---------------12--------------->
  ----------13---------->  <-----------26------------
  ----------14----------->  <-----------28--------------
      --------------15------------------->
       -----------16--------------->
          <-----------------17-----------------
          <-----------------18------------------
          <-----------------19------------------
          <-----------------20------------------
        <--------21--------->
           <-------------22--------------->
```

#	EST#	GB#	ATCC#	left	right	library
1	F	N67699	548297	1	394	Skin
2	F	W03370	548297	1	461	Skin
3	F	H02281	388687	2	267	Placenta
4	F	W60713		8	612	
5	A EST18348	T30547	100338	11	443	Liver
6	B EST136931			11	276	Epididymis
7	F	R52974	364940	15	464	Brain
8	F	R63477	376560	15	428	Placenta
9	L	F09898		18	342	Brain
10	A EST47794		102853	18	435	Spleen
11	F	N52486	503375	18	480	Liver
12	F	N71785	547836	18	432	Skin
13	F	Z39589		18	312	Brain
14	F	T97337	343633	20	347	Liver
15	F	H96364		67	455	
16	L	Z25173		124	457	Skeletal muscle
17	F	T55667	313401	183	620	Spleen
18	F	W57675		183	628	
19	F	H02384	388687	204	643	Placenta
20	F	H66554	480712	204	638	Liver
21	A EST74962		169100	206	449	Brain
22	F	R05588	346947	253	637	Liver
23	A EST75894		169916	318	534	Brain
24	F	H51965	444198	363	793	Liver
25	B EST157610			405	862	Salivary gland
26	F	T49496	307397	407	695	Placenta
27	F	R63476	376560	422	849	Placenta
28	B EST144896			431	752	Brain
29	A EST38789		136236	572	930	Embryo
30	F	W60740		587	986	
31	B EST163071			608	817	Synovial membrane

```
Sequence source codes:
F = WashU/Merck
A = TIGR
B = HGS
L = Genethon
```

FIGURE 1 An example of a Tentative Human Consensus from The Institute for Genomic
Research's Human cDNA Database

(347 ORFs, 19.9% of the total) have significant similarity to hypothetical proteins in other organisms, indicating that they likely represent important, conserved physiological functions.

Since the completion of *H. influenzae*, we have applied the same whole-genome shotgun approach to sequence five additional genomes. *Mycoplasma genitalium*, with 470 genes within 0.58Mb, has the smallest genome of any free-living organism (Fraser *et al.*, 1995). *Methanococcus janaschii*, a deep-sea, hyperbaric, thermophilic methanogen, represents the first sequence of a member of the Archaea (Bult *et al.*, 1996); its gene content clearly demonstrates that the Archaea represent a domain of life distinct from the Eubacteria and Eukaryotes. *Helicobacter pylori* is known to be the causative agent of gastric ulcers and a contributing factor in the development of stomach cancer. The genes identified through sequencing its genome reveal the mechanisms by which it can carry out the unique metabolic functions required for it to survive the diverse, hostile environments it inhabits (Tomb *et al.*, 1997). *Archaeoglobus fulgidus*, a source of deep-sea oil well souring, is the second representative of the Archaea and the first sulfate-reducing organism to be sequenced and adds significantly to our knowledge of the archaean domain (Klenk *et al.*, 1997). *Borrelia burgdorferi*, which causes Lyme disease, is the first spirochete to be sequenced. Spirochetes have been suggested as prokaryotic precursors of the sensory perception machinery in animals (Margulis, 1996) and as such may provide important evolutionary insight into the evolution of eukaryotic cells (Fraser *et al.*, 1997). Representations of the genomes completed at TIGR are shown in Figure 2 and a list of completed and ongoing prokaryotic sequencing projects at TIGR is included in Table 1.

The sequence of these microbial genomes represents the first crucial step in understanding the physiology of these organisms. By analyzing their gene content and supplementing it with existing experimental evidence, we have gained great insight into how these organisms sense, adapt to, and exploit their environments. This has allowed us to reconstruct many of their metabolic pathways and, where known essential pathways appear incomplete, to postulate mechanisms for other genes that may play crucial roles. However, as nearly 50% of the genes identified in these prokaryotes cannot be assigned a function based on sequence similarity to proteins of known function, it is an indication of how much we have yet to learn. What is most exciting and encouraging is the large fraction of conserved hypothetical proteins that have homologues in multiple species, indicating that they are likely play some important, but as yet undiscovered physiological role.

EUKARYOTIC GENOME SEQUENCING: ADDRESSING HUMAN HEALTH AND SUSTAINABILITY

Malaria is the most important tropical parasitic disease in the world. Nearly 40% of the world's population lives in malaria-endemic areas and the disease afflicts 300-500 million people and causes 1.5-2.6 million deaths annually. During the past 20 years, the increase of parasite resistance to drugs, the increase of mosquito resistance to insecticides, and the deterioration of the infrastructure necessary to maintain health programs for controlling the

TABLE 1 Microbial genome projects completed at TIGR. An additional 40 microbial projects are planned or underway; for a current list, see:
<http://www.tigr.org/tdb/mdb/mdb.html>

Organism	Status	Size (Mb)	# ORFs
Haemophilus influenzae	Science 269: 496, 1995	1.83	1741
Mycoplasma genitalium	Science 270: 397, 1995	0.58	470
Methnococcus jannaschii	Science 273: 1017, 1996	1.7	1737
Archaeoglobus fulgidus	Nature, in press	2.2	2436
Borrellia burgdorferii	Nature, in press	1.3	1283
Treponema pallidum	complete, manuscript in preparation	1.0	~950

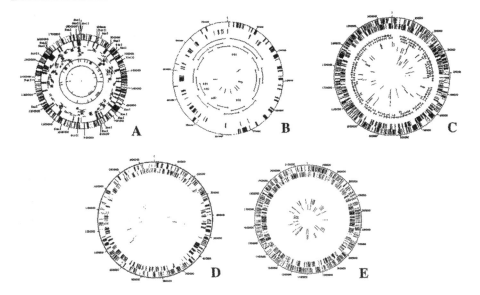

FIGURE 2   Microbial genomes completely sequenced at TIGR: **A** *Haemophilus influenzae*, **B** *Mycoplasma genitalium*, **C** *Methanococcus jannaschii*, **D** *Helicobacter pylori*, **E** *Archaeoglobus fulgidus*.

disease have resulted in a resurgence of malaria worldwide. If predictions of continued global climate change hold true, one million additional fatalities per year may occur in tropical and temperate regions by the middle of the next century (Patz *et al.*, 1996). Consequently, there is an urgent need to develop new drugs and vaccines for the prevention and treatment of malaria.

Unfortunately, development of drugs and vaccines against malaria has proceeded much too slowly. The parasite has a complex life cycle involving both vertebrate and invertebrate hosts, and only the erythrocytic stage can be cultured. Access to parasite macromolecules expressed in other parts of the life cycle is therefore severely restricted, limiting our understanding of parasite biology. The difficult, time-consuming, and expensive process of identifying the target molecules in the parasite has impeded development of drugs and vaccines against malaria. Clearly, new strategies are needed to understand the biology of the parasite and to identify key targets for vaccine and drug development. The Malaria Genome Project, an international effort recently established to sequence the entire genome of *Plasmodium falciparum*, the most deadly human malaria parasite, is an important step in this direction.

Building on our success in the sequencing of microbial genomes, TIGR has adopted a whole-chromosome shotgun approach to sequencing chromosomes 2, 10, 11, and 14 of this eukaryotic pathogen. Chromosome 2, nearly 1 Mb in size, should be completed by the end of 1997 and as such will be the first complete eukaryotic chromosome completed outside of yeast. TIGR is participating in this project as part of a collaborative effort involving a number of groups around the world that should allow the entirety of the 30 Mb genome to be sequenced within the next five years. The ultimate goal of this project is to identify the estimated 10,000 genes encoded within the parasite genome and to provide the complete nucleotide sequence of every potential target molecule for drug and vaccine development. The sequence information and reagents developed as part of this project will lead to a better understanding of the biology of *P. falciparum* and its interactions with its hosts. New strategies for disease control and prevention will certainly follow.

Increasing human population must be met by increasing crop production. Plant biologists have long realized the value of genetic information and have become adept at breeding plants with improved yields, drought tolerance, and parasite resistance. However, the success of microbial genome sequencing projects in identifying crucial genes and pathways has led to the realization that plant genomic sequence can help to significantly accelerate the process. *Arabidopsis thaliana* has rapidly become the model organism for plant biologists who aim to use genes from the small crucifer to tackle larger and more economically important crop plants.

Our approach to sequencing *Arabidopsis* relies on shotgun sequencing of bacterial artificial chromosome (BAC) clones and the identification of new sequencing targets using an innovative BAC end sequencing strategy that eliminates the need for mapping before sequencing (Venter *et al.*, 1996). BAC clones representing a 10x coverage of the *A. thaliana* genome are being sequenced from both ends, producing a sequence tag, on average, every 5 kb across the genome. This collection of sequence data represents the ideal resource for identifying minimal overlapping clones for sequencing. Once a BAC has been completely sequenced, one only has to search it against the BAC end database to identify, by sequence identity, the overlapping BAC whose end is closest to that of the completed BAC. This approach guarantees the smallest amount of redundant sequencing possible while avoiding the need for clone mapping.

To date, we have completed nearly 2 Mb of *A. thaliana* genomic sequence and an additional 3 Mb are in progress. With a density of one gene per 5 kb, the 10 Mb of sequence being produced over the next few years by the TIGR *Arabidopsis* sequencing project should yield sequence information on over 2,000 plant genes. The collaborative effort of *Arabidopsis* sequencing groups worldwide should complete 50% of the 100 Mb genome in the next three years. The early success of this project should allow us to rapidly complete the sequencing of the *A. thaliana* genome, providing us with nearly 20,000 plant genes for further study and characterization.

SEQUENCING THE HUMAN GENOME

Sequencing the genomes of prokaryotes and other eukaryotes will provide invaluable data for the understanding of gene functions and the metabolic capability necessary to sustain life. However, the ultimate goal of genome sequencing is to sequence the human genome. Groups in Japan, Europe, and across the US are currently working toward the goal of completing a reference human genome sequence by 2005. TIGR is an active participant in that effort and we expect to continue to be a significant contributor to this important endeavour.

TIGR has been awarded a grant from the National Human Genome Research Institute to sequence the short arm of human chromosome 16. As one of only six NIH-funded large-scale pilot Human Sequencing Centers, we are committed to completing nearly 30 Mb of human genomic sequence before March 1999. To date, we have completed nearly 4.5 Mb of genomic sequence from chromosome 16, and we have an additional 8 Mb in progress. As with *Arabidopsis*, we are sequencing BAC clones and are using the BAC end strategy for selecting overlapping clones for sequencing. In support of this project, we have worked to develop both laboratory protocols and software to facilitate sample management, quality control, closure, finishing, and annotation of BAC projects to produce the highest quality finished sequence data possible. We anticipate that human genomic sequencing will scale to a rate of nearly 100 Mb per year.

Without an investment in the development of robust computational tools to support sequencing, such large projects would be impossible. Of particular importance is the development of tools for genomic sequence annotation. Without accurate and reliable annotation of genes and features, genomic sequence data are of limited value. Consequently, we have devoted a significant effort to developing tools to streamline the automation process. We have developed a set of visualization and editing tools that allow a trained annotator, starting from a combination of gene prediction programs and searches against EST, THC, and protein databases, to build an accurate gene model quickly and both to determine appropriate gene structure and to assign a putative gene function. Through extensive experience, we have found that automated methods, without the interpretation that a trained biologist can provide, are of limited utility. By automating much of the annotation process, we have constructed a system that allows such skilled individuals to annotate large genomic regions efficiently and accurately.

THE NEXT GENOME PROJECT: FROM SEQUENCE TO FUNCTION

While extremely valuable, the collection of sequence data represents only the first step in developing an understanding of fundamental processes underlying life. Across species, nearly half of the candidate genes that have been identified cannot be assigned a biological role based on similarity to proteins of known function. Further, one should even consider genes with putative functional assignments to be hypothetical until experimental evidence, such as an actual protein product, is presented. In this context, a predicted protein sequence generated by a genome sequencing project is simply a starting point in the search to determine how the gene product actually functions within the cell. Determination of the function of each protein expressed in a cell or organism is clearly the target for the next generation of whole-genome analysis, "functional genomics."

In order to keep pace with the volume of sequence data, functional genomics approaches need to be amenable to high-throughput analysis of input samples. Recently, a number of techniques have been developed that allow the monitoring of relative expression of mRNA messages in different cell types or stages. One of the most promising is microarray analysis. The power of this technique is its potential to allow the monitoring of gene expression through the measurement of mRNA levels in cells cultured under a variety of conditions, by simultaneously interrogating all of the genes within an organism.

Microarray expression analysis is based on the construction of ultra-high density "microarrays" of cDNA or ORFs on glass microscope slides, followed by hybridization with fluorescently labeled cDNA and analysis using a confocal laser scanner. The utility of this approach is two-fold. First, the microarrays allow the simultaneous interrogation of thousands of genes with mRNA from the tissue or cell type of interest. The target cDNA clones can be intelligently chosen for inclusion in the arrays using the vast cDNA and EST resources that have been generated in the past few years. Second, the use of fluorescently labeled probes and confocal laser microscopy allows quantitative determination of the relative expression levels of thousands of genes in a single experiment.

This technology has been developed independently by two groups. Patrick Brown and collaborators at Stanford have developed a prototype instrument that has proven to be extremely useful (Schen et al., 1995; DeRisi et al., 1996). At the same time, Molecular Dynamics Corporation (MD) of Sunnyvale, California began development of a similar arrayer and scanner with the goal of producing a commercially-available instrument. They have formed a partnership with Amersham Life Science, Inc., to develop robust and inexpensive reagents for microarray construction, probe labeling, and hybridization. TIGR has recently entered into collaboration with MD and Amersham to accelerate the development of this technology and to make it a useful and accessible laboratory technique.

We are currently applying this technique to the study of gene expression in prokaryotes, both *Archaea* and *Eubacteria*. The advantage of using prokaryotic systems is that all of the genes from each organism can be arrayed on a single microscope slide. By analyzing expression under a variety of conditions and looking at correlations between gene expression patterns, we will not only gain information relevant to the function of individual genes, but we will also begin to glimpse the transcriptional networks that underlie cellular metabolism. It is the elucidation of these networks that holds the greatest promise for providing us with clues as to the fundamental processes necessary for life.

Analysis of expression in eukaryotes is also progressing. Starting from our THC assemblies, we are identifying cDNA clones, amplifying their inserts, and arraying them with the goal of producing a 30,000 gene human chip. Using this as our starting point, we will analyze gene expression patterns in a variety of human cell types and tissues, focusing on identifying gene expression patterns that are diagnostic of human cancer stage and progression. This should lead not only to a better understanding of the molecular mechanisms underlying human cancer development, but also provide an expression fingerprint that, when combined with clinical data, may allow the selection of the most appropriate and effective treatment regime.

These projects simply represent a first attempt to understand gene function. The challenge for the future will be to develop and refine these techniques and integrate their results with existing data in a well developed database that will allow us to develop a picture of cellular metabolism and gene function. As is true of DNA sequencing, TIGR aims to be one of the leaders in the development and application of these new genomic approaches to address questions of fundamental biological significance.

REFERENCES

Adams, M.D., Kelley, J.M., Gocayne, J.D., Dubnick, M., Polymeropoulos, M.H.M., Xiao, H., Merril, C.R., Wu, A., Olde, B., Moreno, R.F., Kerlavage, A.R., McCombie, W.R., and Venter, J.C. (1991) Complementary DNA sequencing: expressed sequence tags and human genome project. *Science* **252**, 1651-1656.

Adams, M.D., Kerlavage, A.R., Fleischmann, R.D., Fuldner, R.A., Bult, C.J., Lee, N.H., Kirkness, E.F., Weinstock, K.G., Gocayne, J.D., White, O., Sutton, G., Blake, J.A., Brandon, R.C., Chiu, M.-W., Clayton, R.A., Cline, R.T., Cotton, M.D., Earle-Hughes, J., Fine, L.D., FitzGerald, L.M., FitzHugh, W.M., Fritchman, J.L., Geoghagen, N.S.M., Glodek, A., Gnehm, C.L., Hanna, M.C., Hedblom, E., Hinkle, P.S. Jr., Kelley, J.M., Klimek, K.M., Kelley, J.C., Liu, L.-I., Marmaros, S., Merrick, J.M., Moreno-Palanques, R.F., McDonald, L.A., Nguyen, D.T., Pellegrino, S.M., Phillips, C.A., Ryder, S.E., Scott, J.L., Saudek, D.M., Shirley, R., Small, K.V., Spriggs, T.A., Utterback, T.R., Weidman, J.F., Yi, L., Barthlow, R., Bednarik, D.P., Cao, L., Cepeda, M.A., Coleman, T.A., Collins, E.-J., Dimke, D., Feng, P., Ferrie, A., Fischer, C., Hastings, G.A., He, W.-W., Hu, J.-S., Huddleston, K.A., Greene, J.M., Gruber, J., Hudson, P., Kim, A., Kozak, D.L., Kunsch, C., Ji, H., Haodong, L., Meissner, P.S, Olsen, H., Raymond, L., Wei, Y.-F., Wing, J., Xu, C., Yu, G.-L., Ruben, S.M., Dillon, P.J., Fannon, M.R., Rosen, C.A., Haseltine, W.A., Fields, C., Fraser, C.M. and Venter, J.C. (1995) Initial assessment of human gene diversity and expression patterns based upon 83 million nucleotides of cDNA sequence. *Nature* **377 (Supp)**, 3-174.

Affara, N.A., Bentley, E., Davey, P., Pelmear, A. and Jones, M.H. (1994) The identification of novel gene sequences of the human adult testis. *Genomics* **22**, 205-210.

Bult, C.J., White, O., Olsen, G.J., Zhou, L, Fleischmann, R.D., Sutton, G.G., Blake, J.A., FitzGerald, L.M., Clayton, R.A., Gocayne, J.D., Kerlavage, A.R., Dougherty, B.A., Tomb, J.-F., Adams, M.D., Reich, C.I., Overbeek, R., Kirkness, E.F., Weinstock, K.G., Merrick, J.M., Glodek, A., Scott, J.L., Geoghagen, N.S.M., Weidman, J.F., Fuhrmann, J.L., Nguyen, D., Utterback, T.R., Kelley, J.M., Peterson, J.D., Sadow, P.W., Hanna, M.C., Cotton, M.D., Roberts, K.M., Hurst, H.A., Kaine, B.P., Borodovsky, M., Klenk, H.-P., Fraser, C.M., Smith, H.O., Woese, C.R. and Venter, J.C. (1996) Complete genome sequence of the methanogenic archaeon, *Methanococcus jannaschii*. *Science* **273**, 1058-1073.

Collins, F. and Galas, D. (1993) A new five-year plan for the U.S. Human Genome Project. *Science* **262**, 43-46.

DeRisi, J.D., Denland, L., Brown, P.O., Bittner, M.L., Meltzer, P.S., Ray, M., Chen, Y., Su, Y.A. and Trent, J.M. (1996) Use of a cDNA microarray to analyse gene expression patterns in human cancer. *Nature Genetics* **14**, 457-460.

Fleischmann, R.D., Adams, M.D., White, O., Clayton, R.A., Kirkness, E.F., Kerlavage, A.R., Bult, C.J., Tomb, J.-F., Dougherty, B.A., Merrick, J.M., McKenney, K., Sutton, G., FitzHugh, W., Fields, C., Gocayne, J.D., Scott, J., Shirley, R., Liu, L.-I., Glodek, A., Kelley, J.M., Weidman, J.F., Phillips, C.A., Spriggs, T., Hedlom, E., Cotton, M.D., Utterback, T.R., Hanna, M.C., Nguyen, D.T., Saudek, D.M., Brandon, R.C., Fine, L.D., Fritchman, J.L., Fuhrmann, J.L., Geoghagen, N.S.M., Gnehm, L.A., McDonald, L.A., Small, K.V., Fraser, C.M., Smith, H.O. and Venter, J.C. (1995) Whole-genome random shotgun sequencing and assembly of *Haemophilus influenzae* Rd. Science **269**, 496-512.

Fraser, C.M., Gocayne, J.D., White, O., Adams, M.D., Clayton, R.A., Fleischmann, R.D., Bult, C.J., Kerlavage, A.R., Sutton, G., Kelley, J.K., Fritchman, J.L., Weidman, J.F., Small, K.V., Sandusky, M., Fuhrmann, J.L., Nguyen, D., Utterback, T.R., Saudek, D.M., Phillips, C.A., Merrick, J.M., Tomb, J.-F., Dougherty, B.A., Bott, K.F., Hu, P.-C., Lucier, T.S., Peterson, S.N., Smith, H.O., Hutchison, C.A. III and Venter, J.C. (1995) The *Mycoplasma genitalium* genome sequence reveals a minimal gene complement. *Science* **270**, 397-403.

Fraser, C.M., Casjens, S., Huang, W.M., Sutton, G.G., Clayton, R.A., Lathigra, R., White, O., R Dodson, R.J., Hickey, E.K., Gwinn, M., Dougherty, B.A., Ketchum, K.A., Tomb, J.-F., Fleischmann, R.D., Richardson, D., Kerlavage, A.R., Quackenbush, J., Salzberg, S., Hanson, H., van Vugt, R., Palmer, N., Peterson, J., Gocayne, J.D., Weidman, J.F., Utterback, T., Watthey, L., McDonald, L., Artiach, P., Bowman, C., Garland, S.A., Fujii,

C., Cotton, M.D., Horst, K., Roberts, K., Hatch, B., Smith, H.O. and Venter, J.C. (1997) Sequence analysis of the multi-element genome of *Borrelia burgdorferi*, the Lyme Disease spirochete. *Nature*, in press.

Klenk, H-P., Clayton, R.A., Tomb, J.-F., White, O., Nelson, K.E., Ketchum, K.A., Dodson, R.J., Gwinn, M., Hickey, E.K., Peterson, J.D., Richardson, D., Kerlavage, A.R., Graham, D.E., Kyrpides, N.C., Fleischmann, R.D., Quackenbush, J., Lee, N.H., Sutton, G.G., Gill, S.R., Kirkness, E.F., Dougherty, B.A., McKenney, K., Adams, M.D., Loftus, B.J., Peterson, S.N., Reich, C.I., McNeil, L.K., Badger, J.H., Glodek, A., Zhou, L., Overbeek, R., Gocayne, J.D., Weidman, J.F., McDonald, L., Utterback, T., Cotton, M.D., Spriggs, T., Artiach, P., Kaine, B.P., Sykes, S.M., Sadow, P.W., D'Andrea, K.P., Bowman, C., Fujii, C., Garland, S.A., Mason, T.M., Olsen, G.J., Fraser, C.M., Smith, H.O., Woese, C.R. and Venter, J.C. (1997) The complete genome sequence of the hyperthermophilic, sulfate-reducing archaeon *Archaeoglobus fulgidus*. *Nature*, in press.

Margulis, L. (1996) Archaeal-eubacterial mergers in the origin of Eukarya: phylogenetic classification of life. *Proceedings of the National Academy of Sciences, USA* **93**, 1071-1076

McCombie, W.R., Adams, M.D., Kelley, J.M., FitzGerald, M.G., Utterback, T.R., Khan, M., Dubnick, M., Kerlavage, A.R., Venter, J.C. and Fields, C. (1992) *Caenorhabditis elegans* expressed sequence tags identify gene families and potential disease gene homologues. *Nature Genetics* **1**, 124-131.

Patz, J.A., Epstein, P.R., Burke, T.A. and Balbus, J.M. (1996) Global climate change and emerging infectious diseases. *JAMA* **275**(3), 217-223.

Schena, M., Shalon, D., Davis, R.W. and Brown, P.O. (1995) Quantitative monitoring of gene expression patterns with complementary DNA microarray. *Science* **270**, 467-470.

Schuler, G.D., Boguski, M.S., Stewart, E.A., Stein, L.D., Gyapay, G., Rice, K., White, R.E., Rodriguez-Tome, P., Aggarwal, A., Bajorek, E., Bentolila, S., Birren, B.B., Butler, A., Castle, A.B., Chiannikulchai, N., Chu, A., Clee, C., Cowles, S., Day, P.J.R., Dibling, T., Drouot, N., Dunham, I., Duprat, S., East, C., Edwards, C., Fan, J.-B., Fang, N., Fizames, C., Garrett, C., Green, L., Hadley, D., Harris, M., Harrison, P., Brady, S., Hicks, A., Holloway, E., Hui, L., Hussain, S., Louis-Dit-Sully, C., Ma, J., MacGlivery, A., Mader, C., Maratukulam, A., Matise, T.C., McKusick, K.B., Morissette, J., Mungall, A., Muselet, D., Nusbaum, H.C., Page, D.C., Peck, A., Perkins, S., Piercy, M., Qin, F., Quackenbush, J., Ranby, S., Reif, T., Rozen, S., Sanders, C., She, X., Silva, J., Slonim, D.K., Soderlund, C., Sun, W.-L., Tabar, P., Thangarajah, T., Vega-Czarny, N., Vollrath, D., Voyticky, S., Wilmer, T., Wu, X., Adams, M.D., Auffray, C., Walter, N.A.R., Brandon, R., Dehejia, A., Goodfellow, P.N., Houlgatte, R., Hudson, J.R., Jr., Ide, S.E., Iorio, K.R., Lee, W.Y., Seki, N., Nagase, T., Ishikawa, K., Nomura, N., Phillips, C., Polymeropoulos, M.H., Sandusky, M., Schmitt, K., Berry, R., Swanson, K., Torres, R., Venter, J.C., Sikela, J.M., Beckman, J.S., Weissenbach, J., Myers, R.M., Cox, D.R.,

James, M.R., Bentley, D., Deloukas, P., Lander, E.S. and Hudson, T.J. (1996) A gene map of the human genome. *Science* **274,** 540-546.

Schuler, G.D. (1997) Pieces of the puzzle: expressed sequence tags and the catalog of human genes. *Journal of Molecular Medicine* **75,** 694-698.

Sutton, G.G., White, O., Adams, M.D. and Kerlavage, A. R. (1995) TIGR Assembler: A new tool for assembling large shotgun sequencing projects. *Gen. Sci. Tech.* **1,** 9-19.

Tomb, J.-F., White, O., Kerlavage, A.R., Clayton, R.A., Sutton, G.G., Fleischmann, R.D., Ketchum, K.A., Klenk, H.P., Gill, S., Dougherty, B.D., Nelson, K., Quackenbush, J., Zhou, L., Kirkness, E.F., Peterson, S., Loftus, B., Richardson, D., Dodson, R., Khalak, H.G., Glodek, A., McKenney, K., Fitzegerald, L.M., Lee, N., Adams, M.D., Hickey, E.K., Berg, D.E., Gocayne, J.D., Utterback, T.R., Peterson, J.D., Kelley, J.M., Cotton, M.D., Weidman, J.M., Fuji, C., Bowman, C., Watthey, L., Wallin, E., Hayes, W.S., Borodovsky, M., Karp, P.D., Smith, H.O., Fraser, C.M. and Venter, J.C. (1997) The complete genome sequence of the gastric pathogen *Helicobacter pylori*, Nature **386** 539-547.

Waterston, R., Martin, C., Craxton, M., Hunyh, C., Coulson, A., Hillier, L., Durbin, R., Green, P., Shownkeen, R., Halloran, N., Metzstein, M., Hawkins, T., Wilson, R., Berks, M., Du, Z., Thomas, K., Thierry-Mieg, J. and Sulston, J. (1992) A survey of expressed genes in *Caenorhabditis elegans*. *Nature Genetics* **1,** 114-123.

Venter, J.C., Smith, H.O. and Hood, L. (1996) A new strategy for genome sequencing. *Nature* **381,** 364-366

COMPARATIVE GENOMICS AND THE SEQUENCING OF THE MOUSE GENOME

S.D.M. BROWN, A.M. MALLON, R. BATE, M. STRIVENS, P. DENNY

MRC Mouse Genome Centre and Mammalian Genetics Unit, Harwell, Oxon OX11 ORD, UK

M.R.M. BOTCHERBY, R. STRAW, G.W. WILLIAMS, S. FERNANDO, Y. UMRANIA, P.M. WOOLLARD, M. GILBERT, K. GOODALL, J.S. GREYSTRONG, M. RHODES, C.R. MUNDY

UK Human Genome Mapping Project Resource Centre, Hinxton Hall, Hinxton, Cambridgeshire CB10 1SB, UK

G.E. HERMAN, A. DANGEL

Department of Pediatrics and Children's Hospital Research Foundation, Ohio State University, Columbus, Ohio 43205, USA

ABSTRACT

The mouse is an important model organism in the Human Genome Project and is set to play a pivotal role in the comparative analysis of diverse genomes that will be increasingly important for studies of gene function as we move post-genomics. In mouse genetics, it is recognised that the systematic generation of new mouse mutations along with identification of the underlying genes is an important challenge for gene function studies in the post-genomics era. Large numbers of mouse mutations affecting a plethora of biological pathways and with a diverse range of phenotypes need to be generated and mapped in the mouse genome. Many mouse mutations will be homologues of known human genetic disease loci and uncovering the underlying gene to any mouse mutation will shed light on gene function both in the mouse, human and other species. Sequencing of the mouse genome and its comparison with human genome sequence will not only aid gene identification but will provide a rapid route to uncovering genes underlying any mouse mutation. Furthermore, the sequence comparison of two mammalian genomes can be expected to provide profound new insights into gene regulation and genome evolution.

INTRODUCTION

Genome sequencing from a vast range of life-forms from Eubacteria to Eukarya is gathering pace (Clayton *et al.*, 1997). A large number of bacterial genomes have now been sequenced, including *E. coli* (Koonin, 1996). Recently, the complete sequence of a eukaryote, *Saccharomyces cerevisiae*, has been reported (Dujon, 1996). The sequencing of the nematode genome, *C. elegans*, is well advanced with an expectation that its sequence will be complete by end 1998. Sequencing of the human genome has begun in earnest, though only a few percent has been completed to date. Of the other mammal that has featured prominently in the genome project - the mouse - little has been heard. Yet there are strong arguments for undertaking the sequencing of the mouse genome.

The mouse is one of our best characterised mammalian model organisms. It is our best genetic tool for the study of mammalian genetics (Deitrich *et al.*, 1995). A total of around 17,000 markers have now been mapped in the mouse, including over 5,000 genes and over 10,000 DNA markers (Graves, 1996). Nearly 2,000 homologous loci have been identified between the mouse and human genomes. These loci form some 113 conserved linkage groups between the mouse and human genomes in which gene content and gene

Genomics

order are conserved. This provides a powerful tool to reciprocally relate genetic information between the two genomes - to identify mutants in mouse that may be homologous to disease loci in human (and vice versa) and to identify genes in the mouse as candidates for human disease loci or, alternatively, identify human genes that may underlie mouse mutations (Brown, 1994).

The burgeoning genetic maps provide an important framework for the construction of physical maps that will ultimately be required for the generation of templates for the sequencing of the mouse genome (Brown, 1996). Considerable effort is currently being placed on generating genome-wide YAC (yeast articifical chromosome) clone contigs of the mouse genome (For summaries of current progress visit the web sites at Whitehead Institute for Genome Research, http://www-genome.mit.edu, and the UK Mouse Genome Centre, http://www.mgc.mrc.ac.uk). YAC clone contigs are the first step in providing the necessary sequence ready maps for any systematic effort towards sequencing regions of the mouse genome (see below).

Although the mouse is our most powerful genetic tool in mammalian genetics, it is increasingly recognised that the mouse mutation resource is limited and that we have access to only a limited number of mutant phenotypes. Currently, there are around 1,000 mouse mutations that represent only a fraction (1-2%) of the total number of mammalian genes. There is a "phenotype gap" (Brown and Peters, 1996) which needs to be filled by systematic mutagenesis programmes. Such an effort is underway to generate a large number of new mouse mutations and place them on the genetic map. In tandem there is a considerable effort underway in both human and mouse to improve the density of the mammalian gene map. Much of this effort is coming via the mapping of Expressed Sequence Tags, ESTs (Schuler *et al.,* 1996). Importantly, given the conserved segment relationships that have been established between the mouse and human genomes, the positioning of new gene sequences on either genome will generally define the position of the new gene in the other species as well. As the mouse mutant map grows and the mammalian gene map develops in parallel we can expect an increasing number of mutant genes to be identified by a positional candidate approach.

Nevertheless, we are some way from a comprehensive gene map and it is understood that the development of a complete human gene map will require access to the sequence of the human genome. At the same time, mouse sequencing will assist in the process of identifying genes in the mammalian genome. Both the mouse and human genome sequence will together be a powerful tool for the exploitation of the growing mouse mutation resource. In addition, comparative mouse and human sequencing can be expected to shed light on the many still mysterious aspects of genome and chromosome evolution. However, few regions of the mouse genome are currently targets for sequencing. Examples to date include the T cell receptor complex (Koop and Hood, 1994) and the *Btk* region of the mouse X chromosome (Oeltjen *et al.,* 1997). However, increasing efforts are being placed on long-range sequencing in the mouse genome covering areas of genic and structural interest - and we describe here our current approaches to the sequencing of a region of the mouse X chromosome.

SEQUENCING THE MOUSE GENOME - THE BENEFITS

It is useful at the outset to outline some of the specific benefits that are likely to accrue from the comparative sequencing of the mouse and human genomes

Gene identification

The identification of genes from large scale genomic sequence is one route for finding the coding regions underlying new mouse mutations and thus an important aspect of future gene function studies. As will be discussed in more detail below, a variety of gene finding algorithms are available to scan genomic sequence from both human and mouse in order to identify potential coding regions. Gene finding algorithms scan for a number of sequence features, but tend to have limitations. The direct comparison of large-scale mouse and human genome sequence to identify conserved sequence regions is expected to be an important adjunct to assessing exons identified using gene prediction software. Moreover, comparison of mouse and human sequences can also be expected to identify regulatory elements flanking and within genes.

Gene and repeat sequence distribution

It will be important to compare both the distribution of genes across large genomic regions in mouse and man, examining both gene order and intergenic distances. While a large number of regions of the mouse and human genomes show conservation of known gene content and order (see above) it is unclear how much local rearrangement of gene order has taken place within these so-called conserved segments. In addition, comparison of repeat sequence density and distribution between the mouse and human genomes will be a valuable addition to our understanding of the dynamics of genome evolution. Furthermore, it will be important to assess gene and repeat sequence distribution in the context of cytogenetic band location on chromosomes. Dark Giemsa-staining or dark bands on chromosomes are generally gene-poor and relatively rich in LINE (long interspersed repeat sequences). This contrasts with the light staining bands which are rich in SINES (short interspersed repeat sequences) and gene-rich (Bernardi, 1995; Bickmore & Sumner, 1989). Long range sequence analysis will also be beneficial for the characterisation of the genomic regions associated with chromosomal band boundaries and the search for specific sequence structures within those regions. This would include the search for PABL (pseudoautosomal boundary-like) sequences already identified in the human genome (Fukagawa *et al.*, 1996).

Chromosome evolution

Comparative sequencing programmes will also shed new light on the mechanisms of chromosome evolution. As discussed above there are many conserved linkage groups between the mouse and human genomes in which gene content and gene order are conserved - so-called conserved segments. The catalogue of conserved segments provides a snapshot of the continuing process of chromosome rearrangement that is occurring in the two lineages since mouse and human separated some 80 million years ago. Little is known of the processes involved which generate these chromosome rearrangements; or of the sequences present at the evolutionary breakpoints that define the current limits of conserved segments on each chromosome. Identification and comparison of sequences at the evolutionary breakpoints between mouse and human can be expected to lead to considerable insight into the underlying mechanisms that lead to chromosome evolution.

SEQUENCING THE MOUSE GENOME - TECHNICAL APPROACHES

Though there is still some way to go towards completing genome-wide physical maps of the mouse genome, many chromosome regions are now covered with robust YAC clone contigs which can form the basis for the construction of sequence-ready maps. As in the human genome, most YAC physical maps are being laid down using an STS (sequence tagged site) content approach (Brown, 1996). A high density of STSs in any region can be used to construct overlapping clone contigs by assessing common STS content. However, YAC clones with inserts of around 1Mb are less than ideal substrates for sequencing. YAC contigs form a framework from which to derive contigs of smaller insert clones e.g. BACs/PACs/P1s (bacterial artificial chromosomes, phagemid artificial chromosomes) or cosmids. These clones with insert sizes ranging from 40kb to 150kb are ideal substrates for shotgun sequencing approaches. Deep BAC/PAC/P1 or cosmid contigs are constructed across any chromosome region making use of the YAC clone contigs and STS resources available. The sequence-ready maps can be validated by fingerprinting techniques that will aid the identification of a minimal tiling path of clones for sequencing as well as assess the integrity of the contig clones. The process of generating sequence-ready maps in the mouse follows a very similar strategy to that undertaken in the human.

The smaller insert BAC/PAC/P1 or cosmid clones are shotgun cloned into M13 and sequenced using a mixture of dye primer and dye terminator chemistry (see Figure 1). For a typical BAC, some 2,500 sequence reads are required. Assembly is achieved employing the software PHRED and PHRAP (Phil Green, personal communication) (Figure 1). Contigs are extended by further sequencing of selected subclones, using a combination of Gap4 (Bonfield *et al.*, 1995) and long gels or primer walks. Final contiguation, if not achieved, is reached by PCR amplification and sequencing of remaining gaps. The finished sequence is analysed by two routes carried out somewhat in parallel. Initially the repeat regions are identified using a combination of RepeatMasker (Smit and Green, unpublished results), Censor (Jurka *et al.*, 1996) and Xblast (Altschul *et al.*, 1990). The masked query sequence is then searched against Genbank and Swiss-Prot using BLASTN, BLASTX and TBLASTX algorithms (Altschul *et al.*, 1990) to identify potential homologies to both DNA and protein sequences. The raw output from the BLAST algorithms is filtered using MSPcrunch and viewed through Blixem (Sonnhammer and Durbin, 1994). Alongside database searching, potential exons are identified using a variety of software. We employ mainly GRAIL II (Uberbacher and Mural, 1991) and the Genefinder suite of programs (Solovyev *et al.*, 1995). Both of these gene prediction packages are known to have individual limitations in terms of exon prediction, however comparison of their output increases their predictive power. Due to their limitations it will be increasingly important to compare sequenced mouse regions with available human sequence data where possible. Initial comparison of sequences will be performed using the interactive dot matrix program, DOTTER (Sonnhammer and Durbin, 1995). Modifications of the current dot plot analysis such as percentage identity plots (Oeltjen *et al*, 1997) and LAPS (Schwartz *et al*, 1991) will be explored to identify patterns of sequence conservation.

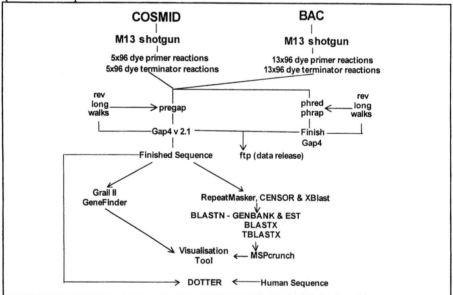

FIGURE 1. Sequencing and analysis strategy showing the process from the initial clones through the sequencing, assembly (phred&phrap), data release and gene identification (GeneFinder and XGrail) in parallel with database searching (BLAST and MSPcrunch) and finally visualisation and comparison (DOTTER).

SEQUENCING THE MOUSE X CHROMOSOME

Our current work is focused on a region of the mouse X chromosome that will provide advances and insights into a number of areas including (see also above):

1. improvements in gene identification in both the human and mouse genomes through the recognition of homologous sequences - allied to this, the identification of candidate genes for mouse mutations and their human disease homologues
2. identification of conserved regulatory sequences
3. detailed analysis of gene order and intergene distance in the two species
4. the structure and nature of evolutionary chromosome breakpoints
5. comparison of repeat sequence and base composition distribution and their relationship to chromosome structure and banding

The *Ids* to *Dmd* region of the chromosome represents around 5Mb of sequence. YAC contigs cover this region aside from two small gaps (Chatterjee *et al.*, 1994; Gouyon *et al.*, 1996). In the mouse this region extends from light band B to dark band C and crosses an evolutionary breakpoint being homologous to two separate conserved segments on the human X chromosome - IDS to F8A [Xq28] and DMD [Xp21.3-21.2] (Herman *et al.*, 1996). In addition, the IDS to F8C region represents a very gene-rich region in keeping with a light band location in both genomes (see above). The adjacent region in the mouse encompasses the large *Dmd* gene which is located in a dark band and represents a gene-poor region. Thus, the *Ids* to *Dmd* region represents an ideal area for sequencing encompassing as it does gene-rich and gene-poor regions that are separated by an evolutionary breakpoint. Finally, this region contains two mouse mutations with interesting disease homologues in the human genome: bare patches (*Bpa*) and striated (*Str*) mutations - possible homologues of X-linked dominant chondrodysplasia punctata (CDPX2) and incontinentia pigmenti (IP2) mutations respectively (Herman *et al.*, 1996). It can be expected that the sequence from the physical region defined as containing *Str* and *Bpa* may aid in gene identification. This region (see Figure 2) which is bounded by the markers DXHXS1104 and DXHXS52 is already covered by extensive cosmid and BAC contigs covering nearly 600kb that are being used to initiate the sequencing effort in this region, while further, more extensive sequence ready maps are being prepared elsewhere. Intriguingly, this region also contains the *F8a* gene which in human is located within the F8C gene (Levinson *et al.*, 1990) and represents the only known exception to the conserved linkage group between the *Ids-Cf8* region in mouse and the IDS-F8C region in human.

SEQUENCE UPDATE

As of end June, 1997 we have accumulated 400kb of sequence from the DXHXS1104 - DXHXS52 region of the mouse X chromosome. Analysis of this sequence is allowing us to identify new genes from this region and to identify potential candidates for the *Bpa* and *Str* mutations. The sequence across the corresponding human region is nearly complete (A. Rosenthal, personal communication) and we expect that comparisons of the human and mouse sequences will assist the identification of further coding regions.

CONCLUSION

The comparative sequencing of the mouse and human genomes is beginning. Comparative sequencing will markedly improve our annotation of the mammalian genome, assisting with gene identification and throwing new light on genome and chromosome evolution.

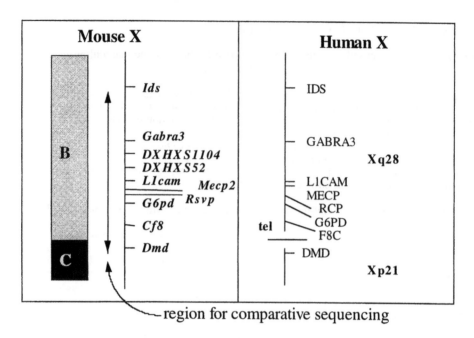

region for comparative sequencing

FIGURE 2. The mouse X region for comparative sequencing is shown on the left with respect to its chromosome banding structure and compared with the homologous regions on the human X (right). The mouse region to be sequenced crosses an evolutionary breakpoint between the mouse and human X chromosomes.

ACKNOWLEDGEMENTS

This work was supported by the Medical Research Council, UK,

REFERENCES

Altschul, S.F., Gish, W., Miller, W., Myers, E.W. and Lipman, D.J. (1990) Basic local alignment search tool. *Journal of Molecular Biology*, **215**, 403-410.

Bernardi, G. (1995) The human genome: Organisation and evolutionary history. *Annual Review of Genetics*, **29**, 445-476.

Bickmore, W. and Sumner, A.T. (1989) Mammalian chromosome banding - an expression of genome organization. *Trends in Genetics*, **5(5)**, 144-148.

Bonfield, J.F., Smith, K.F. and Staden, R. (1995) A new DNA sequence assembly program. *Nucleic Acids Research.* **24**, 4992-4999.

Brown, S.D.M. (1994) Integrating maps of the mouse genome. *Current Opinion in Genetics and Development*, **4**, 389-394

Brown, S.D.M. (1996) Mouse Genome. In: *Encyclopedia of Molecular Biology and Molecular Medicine* Vol. 4. (Ed. R.A. Meyers). VCH Publishers, New York. pp 120 128.

Brown, S.D.M. and Peters. J. (1996) Combining mutagenesis and genomics in the

34

mouse - closing the phenotype gap. *Trends in Genetics*, **12**, 433-435.

Chatterjee, A., Faust, C.J., Molinari-Storey, L., Kiochis, P., Poustka, A. and Herman, G.E. (1994) A 2.3Mb yeast artificial chromosome contig spanning from Gabra3 to G6pd on the mouse X chromosome. *Genomics*, **21**, 49-57.

Clayton, R.A., White, O., Ketchum, K.A. and Venter, J.C. (1997) The first genome from the third domain of life. *Nature*, **387**, 459-462.

Deitrich, W.F., Copeland, N., Gilbert, D.J., Miller, J.C., Jenkins, N.A. and Lander, E.S. (1995) Mapping the mouse genome: current status and future prospects. *Proceedings of the National Academy of Sciences, USA*, **92**, 10849-10853.

Dujon, B. (1996) The yeast genome project: what did we learn? *Trends in Genetics*, **12**, 263-270.

Fukagawa, T., Nakamura, Y., Okumura, K., Nogami, M., Ando, A., Inoko, H., Saitou, N. and Ikemura, T. (1996) Human pseudoautosomal boundary-like sequences: expression and involvement in evolutionary formation of the present-day pseutoautosomal boundary of human sex chromosome. *Human Molecular Genetics*, **5**, 23-32.

Gouyon, B., Chatterjee, A., Monaco, A., Quaderi, N., Brown, S.D.M. and Herman, G.E. (1996) Comparative mapping on the mouse X chromosome defines a myotubular myopathy equivalent region. *Mammalian Genome* , **7**, 575-579.

Graves, J.A.M., (1996) Comparative Genome Organization; First International Workshop. *Mammalian Genome*, **7**, 717-734.

Herman, G., Blair, H.J., Brown, S.D.M., de Gouyon, B., Haynes, A. and Quaderi, N. (1996) The Mouse X Chromosome Committee report. *Mammalian Genome*, **6**, S317 S330.

Jurka, J., Klonowski, P., Dagman, V. and Pelton, P. (1996) CENSOR - a program for identification and elimination of repetitive elements from DNA sequences. *Computers and Chemistry*, **20**, 119-122

Koonin, E.V. (1997) Big time for small genomes. *Genome Research* , **7**, 418-421.

Koop, B.F. and Hood, L. (1994) Striking sequence similarity over almost 100 kilobase of human and mouse T-cell receptor DNA. *Nature Genetics*, **7**, 48-53

Levinson, B., Kenwrick, S., Lakich, D., Hammonds, G. Jr. and Gitschier, J. (1990) A transcribed gene in an intron of the human factor VIII gene. *Genomics* , **7**, 1-11.

Oeltjen, J.C., Malley, T.M., Muzny, D.M., Miller, W., Gibbs, R.A. and Belmont, J.W. Large-scale comparative seqeunce analysis of the human and murine Bruton's Tyrosine Kinase loci reveals conserved regulatory domains. *Genome Research*, **7**, 315-329.

Schuler, G.D., Boguski M.S., Stewart E.A., Stein L.D., Gyapay G., Rice K., White R.E., Rodriguez-Tome P., Aggarwal A., Bajorek E., Bentolila S., Birren B.B., Butler A., Castle A.B., Chiannilkulchai N., Chu A., Clee C., Cowles S., Day P.J., Dibling T., Drouot N., Dunham I., Duprat S., East C., Edwards. C., Fan, J.-B., Fang, N., Fizames, C., Garrett, C., Green, L., Hadley, D., Harris, M., Harrison, P., Brady, S., Hicks, A., Holloway, E., Hui, L., Hussain, S., Louis-Dit-Sully, C., Ma, J., MacGilvery, A., Mader, C., Maratukulam, A., Matise, T.C., McKusick, K.B., Morissette, J., Mungall, A., Muselet, D., Nusbaum, H.C., Page, D.C., Peck, A., Perkins, S., Piercy, M., Qin, F., Quackenbush, J., Ranby, S., Reif, T., Rozen,S.,

Sanders, C., She, X., Silva, J., Slonim, K., Soderlund, G., Sun, W.-L., Tabar, P., Thangarajah, T., Vega-Czarny, N., Vollrath, D., Voyticky, S., Wilmer, T., Wu, X., Adams, M.D., Auffray, C., Walter, N.A.R., Brandon, R., Dehejia, A., Goodfellow, P.N., Houlgatte, R., Hudson Jr, J.R., Ide, S.E., Iorio, R., Lee, W.Y., Seki, N., Nagase, T., Ishikawa, K., Nomura, N., Phillips, C., Polymeropoulos, M.H., Sandusky, M., Schmitt, K., Berry, R., Swanson, K., Torres, R., Venter, J.C., Sikela, J.M., Beckmann, J.S., Weissenbach, J., Myers, R.M., Cox, D.R., James, M.R., Bentley, D., Deloukas, P., Lander, E.S. and Hudson T.J. (1996) A gene map of the human genome. *Science* , **274**, 540-546.

Schwartz, S., Miller, W., Yang, C.M. and Hardison, R.C. (1991) Software tools for analyzing pairwise alignments of long sequences. *Nucleic Acids Research.* **19**, 4663 4667.

Soloveyev, V.V., Salamov, A.A. and Lawrence, C.B. (1995) Identification of human gene structure using linear discriminant functions and dynamic programming. In *Proceedings of the Third International Conference on Intelligent Systems for Molecular Biology* Eds: Rawling, C., Clark, D., Atlman, R., Hunter, L., Lengauer, T., Wodak, S. eds (AAAI Press, Cambridge, England), 367-375.

Sonnhammer, E.L.L. and Durbin, R. (1994) A workbench for large-scale sequence homology analysis. *CABIOS* , **10**, 301-307

Sonnhammer, E.L.L. and Durbin, R. (1995) A dot matrix program with dynamic threshold control suited for genomic DNA and protein analysis. *GENE,* **167**, GC1-10.

Uberbacher, E.C. and Mural, R.J. (1991) Locating protein-coding regions in human DNA sequences by a multiple sensor-neural network approach. *Proceedings of the National Academy of Sciences, USA* **88**, 11261-11265.

THE YEAST GENOME: SYSTEMATIC ANALYSIS OF DNA SEQUENCE AND BIOLOGICAL FUNCTION

STEPHEN G. OLIVER & FRANK BAGANZ

Department of Biomolecular Sciences, UMIST, PO Box 88, Sackville St., Manchester M60 1QD, UK

ABSTRACT

The yeast *Saccharomyces cerevisiae* is the first eukaryotic organism for which a complete genome sequence has become available. This sequence held a number of surprises for yeast researchers, including the total number of genes, the proportion of those genes whose function is completely unknown, and the high level of apparent redundancy within this very small eukaryotic genome. This chapter examines the organisation and evolution of the yeast genome and discusses the various systematic approaches that are being employed to elucidate the function of the approximately 6,000 protein-encoding genes predicted by the sequence.

YEAST AS A MODEL ORGANISM

The yeast, *Saccharomyces cerevisiae*, is one of the most important experimental organisms for the study of eukaryotic molecular genetics. It has a genome size of only 12 Mb (200 times smaller than that of the human genome). This genome is divided into 16 chromosomes which have been defined by both genetical and physical analysis (Mortimer *et al.*, 1992). Despite their small size, yeast chromosomes resemble their counterparts from higher organisms in structure and in their mechanisms of replication, recombination and segregation (reviewed by Newlon, 1988). Furthermore, the pattern of gene expression and genome organisation in yeast is similar to that in higher organisms (Olson, 1991) and a number of genes exhibit both structural and functional homology with genes from multicellular eukaryotes. For example, a cross-species comparison between human and yeast genes by Foury (1997) has shown that for the protein products of 170 human disease-associated genes cloned by function, 52 (31%) share significant amino-acid sequence similarity with yeast gene products. For positionally cloned human genes, the comparable figures are 22 (28%) of 80. The small genome size together with the large chromosome number and a low incidence of repetitive DNA and introns (Oliver *et al.*, 1993) have made *S. cerevisiae* a very suitable subject for systematic DNA sequencing which can provide important information about the genome of higher eukaryotes.

S. cerevisiae exhibits many characteristics which make it an ideal experimental system for the functional analysis of novel genes discovered by systematic DNA sequencing. Firstly, yeast is an unicellular eukaryote which is able to grow on chemically defined media, thus the physiology of the organism can be completely controlled by the investigator. Secondly, *S. cerevisiae* is capable of vegetative growth in both the haploid and diploid state and so the effect of cell type or ploidy on gene action may be determined easily. Finally, and most importantly, excellent tools exist for the genetic manipulation of yeast which permit the precise deletion or replacement of genes whose function is to be investigated (Rothstein, 1983). Because integrative transformation occurs only by homologous recombination, DNA fragments can be introduced into yeast cells and replaced for the wild-type copy of a novel gene with great accuracy. This provides a straightforward approach to associate defined gene mutations (e.g. complete deletions) with biochemical activity *in vitro* and a specific phenotype *in vivo*. Thus, the yeast *S. cerevisiae* is an excellent organism for pioneering the systematic analysis of gene function.

THE YEAST GENOME SEQUENCING PROJECT

The genome of the yeast *S. cerevisiae* has been completely sequenced through an international effort involving about 600 scientists in Europe, North America and Japan and the entire genomic sequence of 12,068 kb was released to public databases in April 1996 (Goffeau *et al.*, 1996). This was an important achievement, not only because it is the first complete nucleotide sequence of a eukaryote and the largest genome sequenced so far, but also due to the role of yeast as an important model organism. The sequence analysis revealed the presence of 6275 open reading frames (ORFs) of which 5885 are potential protein-encoding genes, indicating the compactness of the yeast genome, with almost 70% of the total sequence consisting of ORFs (Dujon, 1996).

In addition, the genome contains approximately 140 ribosomal RNA genes in a large tandem array on chromosome XII, and some 40 genes encoding small nuclear RNAs, and 275 transfer RNA genes, which are distributed throughout the 16 chromosomes. The comprehensive analysis of the yeast genome by the Martinsrieder Institut für Protein Sequenzen (MIPS) has shown that currently only 43% of all potential protein-encoding genes are classified as "functionally characterised" by having genetically and biochemically well-investigated properties (ca. 30%), being members of well-defined protein families, or showing strong homology to proteins with known biochemical function. For ca. 20% of the remaining ORFs, the available data provide only some indication to the biological role of their encoded proteins. Whereas the remaining 38% either show similarities to other uncharacterised proteins or no similarity at all (Mewes *et al.*, 1997). Likewise, similar high proportions of genes with unknown function have been found by systematic sequencing of the bacterial genomes of *Escherichia coli* (Blattner *et al.*, 1997) and *Bacillus subtilis* (Kunst *et al.*, 1997). Furthermore, the same is true for the nematode worm *Caenorhabditis elegans* (Waterson & Sulston, 1995) and the higher plant *Arabidopsis thaliana* (Delseny *et al.*, 1997). Therefore, one might ask, are these genes real and why have they not been discovered previously? It may be that many of these genes are required to deal with physiological challenges that are not encountered in the laboratory, but which can be found in the natural environment of the organism, e.g. in the rotting fig or on the grape (Mortimer & Johnston, 1986) in the case of yeast. Another explanation for this phenomenon could be genetic redundancy.

THE PROBLEM OF REDUNDANCY

A surprising finding was that a large proportion of the 6000 ORFs appeared to be redundant, since identical, or very similar, copies were found elsewhere in the genome (Goffeau *et al.*, 1996, 1997). A comprehensive analysis of the genome identified 53 regions of clustered gene duplications (chromosome homology regions or CHRs; Mewes *et al.*, 1997), suggesting that it might have undergone a complete duplication during its evolutionary history and been reduced to its present size by a number of reciprocal translocation events (Wolfe and Shields, 1997; Keogh *et al.*, 1997). On general evolutionary grounds, one would expect much of this redundancy to be more apparent than real. An example of this distinction is provided by the 15 kb CHR found on chromosomes XIV and III (Lalo *et al.*, 1993). Both regions contain a citrate synthase gene at equivalent positions; however, *CIT2* on chromosome III encodes a peroxisomal enzyme, whereas *CIT1* on chromosome XIV specifies a mitochondrial enzyme. The discovery of a third citrate synthase gene (*CIT3* on chromosome XVI; Jia *et al.*, 1997) indicated that the situation is even more complicated.

The high level of redundancy in the genome might explain the high proportion of new genes, revealed by the complete *S. cerevisiae* genome sequence, that had eluded discovery by conventional or "function first" (Oliver, 1996a) genetics. If these genes are located in the duplicated regions of the genome, it may have been impossible to identify them by classical genetics since single-gene mutations would be without phenotypic impact. In fact, the reverse is true. 16.2 % of genes with experimentally determined functions (that is, genes with names) are found within the CHRs, whereas only 10.1 % of genes of, so far, undetermined function lie within the duplicated portion of the genome. This surprising finding can, perhaps, best be viewed in terms of Wolfe's model for the evolution of the yeast genome (Wolfe & Shields, 1997). Following the complete duplication of the genome, it may have been that there was competition between CHRs for survival as the genome reduced in size, with those chromosome segments that included a gene that had evolved a new, and selectively advantageous, function being retained. Since genes with such evolutionarily favourable functions would be expected to display an overt phenotype upon mutation, they would be discovered preferentially by classical methods.

Whatever the origins and meaning of redundancy within the yeast genome, it is evident that the elucidation of the functions of the novel genes revealed by the sequence presents a major conceptual and experimental challenge. In order to meet this challenge, new and systematic methods of analysis will be required. In this respect, Oliver (1996a) has proposed a taxonomic system which permits the hierarchical organisation of gene function. This system must be hierarchical in order to limit the number of tests which will have to be performed to elucidate the function of any given novel gene in the *S. cerevisiae* genome. There are several approaches which can be applied to achieve this, using both informatic and experimental analyses.

METHODS TO DETERMINE THE FUNCTION OF NOVEL GENES

Bioinformatic analysis

The first step in elucidating the function of a novel gene is to compare the amino acid sequence of its predicted product with the sequences already present in the international data libraries to search for similarities to proteins of known function. In a review by Bork and colleagues (1994), it was shown that homology- and analogy-based predictions of gene functions have proved to be extremely powerful in exploring genomic information. By comparing the output of several genome sequencing projects, an identification of partial function was possible for 40-65% of all predicted proteins. However, these assignments often provide only general descriptions of the predicted protein products, such as 'protein kinase' or 'transcription factor', but give no indication as to their biological role (Oliver, 1996a).

Even when a novel gene shows significant similarity to a sequenced gene of known function, the result may simply reveal that our understanding of the true function is very superficial. For instance, YCL017c which is one of the few essential genes on yeast chromosome III shows greater than 40% amino-acid sequence identity with the NifS protein of nitrogen-fixing bacteria (Oliver *et al.*, 1992). Other NifS homologues have now been found in *Escherichia coli*, *Lactobacillus bulgaricus*, and *Bacillus subtilis* (Leong-Morgenthaler *et al.*, 1994; Sun & Setlow, 1993; Mehta & Christen, 1993). Neither yeast, nor any of these bacteria, fix molecular nitrogen, and both sequence alignments by Ouzunis and Sander (1993) and some experimental data by Leong-Morgenthaler and colleagues (1993) suggest that the NifS proteins are pyridoxal-dependent aminotransferases. In nitrogen-fixing bacteria, studies with *Azotobacter vinelandii* by Zheng and co-workers (1993) revealed that the NifS protein is responsible for inserting sulphur into the centre of the nitrogenase, using pyridoxal phosphate as a co-factor. Thus the discovery of homologous genes in several organisms has stimulated experiments which provided a much more precise description of the protein function than was previously known. This example should illustrate that, although sequence analysis can provide important information and help to establish priorities for the experimental analysis of gene function, it is not a substitute for such experiments.

Generation of specific deletion mutants

In order to determine the function of the 2000 or more genes of completely unknown function using a systematic and hierarchical approach, a large European research network, called EUROFAN (for European Functional Analysis Network) has been established (Oliver, 1996b). Parallel activities are under way in Canada, Japan, and the USA. The first aim of these functional analysis projects is the production of a complete set of 6000 single-gene deletion mutants covering all of the open reading-frames (ORFs) revealed by the complete *Saccharomyces cerevisiae* genome sequence. Such a mutant collection would be a major resource not only for the functional analysis of the yeast genome but should also permit the 'functional mapping' of the genomes of higher organisms onto that of yeast through transcomplementation experiments (Oliver, 1997 a & b).

These deletions are being generated by a PCR-mediated gene replacement method which relies on the high efficiency and accuracy of mitotic recombination in *S. cerevisiae*. This technique was first described for PCR-generated DNA molecules consisting of the
S. cerevisiae HIS3 gene flanked by 30-50 bp of DNA homologous to the target locus (Baudin *et al.*, 1993). However, the use of homologous nutritional markers (e.g. *HIS3* or *TRP1*) in gene replacement experiments resulted in a rather low efficiency of correct integration, particularly in strains which carry only a small deletion of the marker locus (Baudin-Baillieu *et al.*, 1997). Large improvements were achieved when kanMX, a completely heterologous dominant resistance marker, was used as a selectable module (Wach *et al.*, 1994). The kanMX module consists of the coding sequence of the *kan*[r] gene of the bacterial transposon Tn903, encoding aminoglycoside phosphotransferase (Oka *et al.*, 1981) which is expressed in yeast by the use of promoter and terminator sequences from the *TEF* gene of the filamentous ascomycete fungus, *Ashbya gossypii* (Steiner & Philippsen, 1994). Aminoglycoside phosphotransferase activity renders *S. cerevisiae* resistant to the drug geneticin (Jiminez & Davies, 1980). Due to the heterology of the kanMX module, correct disruption of target ORFs with PCR-generated kanMX cassettes containing 35 bp flanking extensions, homologous to yeast chromosomal DNA, occurred with a very low level of illegitimate integration (Wach *et al.*, 1994). Successful targeting of these PCR products is dependent on the perfect homology between the short flanking regions and the target locus. In order to permit gene replacement in any *S. cerevisiae* strain, including those used in industrial processes, an alternative PCR method was developed (Amberg *et al.*, 1995; Wach, 1996). In this technique, a gene disruption cassette with long flanking homology regions is generated in a nested PCR reaction.

Because of the redundancy problem (see above), and in order to study gene interactions, it is necessary to create strains with multiple deletions. Therefore, new vectors have been constructed that permit the re-use of the selectable marker through *in vivo* excision (Längle-Rouault & Jacobs, 1995; Güldener *et al.*, 1996) in addition to new heterologous markers (Wach *et. al.*, 1997). A further development of the PCR-targeting strategy is the generation of deletion mutants that are labelled with unique oligonucleotide tags, called 'molecular bar codes' (Shoemaker *et al.*, 1996) that can be detected by hybridisation to a high-density oligonucleotide array (Schena *et al.*, 1994). This method should greatly accelerate the systematic analysis of the deletion strains on a genome-wide scale, which will be carried out at the level of gene expression and protein synthesis. Furthermore, these deletion mutants will be a valuable resource for the qualitative and quantitative analysis of phenotype.

Systematic qualitative phenotypic analysis

As a precursor to the EUROFAN project, Slonimski and colleagues have developed a large-scale screening method for the systematic phenotypic analysis of single deletion mutants (Rieger *et al.*, 1997). For this purpose, they generated about 80 haploid deletants by PCR-mediated gene replacement (see above) using *HIS3* or *TRP1* as selectable markers and developed some 150 different growth conditions ranging from different carbon and nitrogen sources, through stress conditions (e.g. high temperature), to specific inhibitors and drugs acting on various cellular processes. In order to perform this phenotypic analysis on a large scale, microtiter plates were used for this screening, which should allow the analysis of hundreds of deletion mutants in parallel. So far, they have tested 73 deletion mutants in about 60 different growth conditions and found some phenotype for 62%. Of these, 37% were clear phenotypes (e.g. no growth in the presence of an inhibitor or on a non-fermentable carbon source and lethals), 25% were 'suggestive', whereas no phenotype was found for the remaining 38% of these mutants. However, these figures should be considered as minimal, since not all mutants were tested in all conditions. Furthermore, ca. 30% of the mutants showed several phenotypes, indicating that many single-gene deletions have pleiotropic effects. The results suggest that this simple and inexpensive screening method is a useful tool for the systematic search of biochemical and physiological function of novel genes, particularly if it can be adapted to high throughput screening by automation.

Transcriptome analysis

A powerful tool in the elucidation of gene function is the quantitative analysis of expression levels of a novel gene under a variety of physiological conditions compared to the expression patterns of functionally characterised genes. One approach to the efficient analysis of gene expression on a genome-wide scale is the use of SAGE (serial analysis of gene expression) technology (Velculescu *et al.*, 1995), which was employed by Velculescu and co-workers (1996) to determine the complete set of yeast genes expressed under a given set of conditions (the 'transcriptome'). In their initial studies, these workers were able to detect 4665 genes (ca. 75% of the predicted protein-encoding ORFs) with most genes being expressed at a low level. Kal (1997) has used the SAGE technology to compare yeast gene transcription in cells grown on oleate and cells grown on glucose. Peroxisomal activity is induced by growth on oleate and the 35 genes showing the highest oleate:glucose transcript ratio included many known to encode peroxisomal proteins, but 15 novel genes were also identified as being likely to be associated with peroxisomal metabolism.

An alternative approach to SAGE is the use of hybridisation-array technologies (Schena *et al.*, 1996; Lockhart *et al.*, 1996) for the high-throughput analysis of the transcriptome. DeRisi and co-workers (1997) have applied DNA microarrays, containing almost every yeast ORF, to carry out a comprehensive study of gene expression during the metabolic shift from fermentation to respiration on a genomic scale. They found marked changes in the global pattern of gene expression for ca. 35% of the ca. 6000 putative protein-encoding genes as glucose was progressively depleted from the medium. About half of these differentially expressed genes were previously uncharacterised, thus the response to the diauxic shift provided the first small clues to their possible roles. Whereas the expression profiles observed for genes with known functions provided insights into the cell's response to a changing environment. For example, several classes of genes, such as those involved in the TCA/glyoxylate cycle and carbohydrate storage, were co-ordinately induced by glucose exhaustion, whereas genes involved in protein synthesis, exhibited a decrease in expression. Furthermore, sets of genes could be grouped on the basis of similarities in their expression patterns. The previously characterised members of each of these groups also shared important similarities in their functions and, in most cases, common regulatory mechanisms could be inferred. Moreover, the authors showed that the same DNA microarrays may be used to identify genes whose expression was affected by deletion of the gene encoding the transcriptional co-repressor Tup1p or overexpression of the transcriptional activator gene, *YAP1*. Their results demonstrated the feasibility and utility of the DNA microarray methodology to the analysis of the transcriptome and indicate that this method should enable the classification of all novel genes according to their expression pattern, thus providing clues to their functions.

Analysis of gene expression using gene fusions

Burns *et al.* (1994) have developed a large-scale screening method for analysing gene expression, protein localisation, and disruption phenotypes of many novel ORFs in *S. cerevisiae*. In this approach, a transposon mutagenesis scheme was used to generate a bank of yeast strains, each with a *lacZ* insertion at a random genomic location. They identified 2800 transformants expressing fusion genes during vegetative growth and 55 transformants containing meiotically induced fusion genes. Based on this result, they estimated that 80-86% of the yeast genome contains open reading frames expressed during vegetative growth and that there are 95-135 meiotically induced genes. Other findings concerned the location of some of the fusion proteins. Out of 2373 strains carrying fusion genes analysed, only 245 (10%) fusion proteins localised to discrete subcellular locations whereas the remainder gave just general cytoplasmic staining (58%) or no staining at all (32%). Furthermore, the phenotypes of ca. 150 single-gene disruptants analysed so far showed that most vegetative genes were not essential but ca. 45% of the mutants exhibited growth defects under one of several conditions tested (Ross-MacDonald *et al.*, 1997). This information, together with the results from the expression and localisation of the fusion proteins, may provide useful clues to unravel the function of novel genes.

In a complementary approach, Dang *et al.* (1994) have developed a method to identify yeast genes regulated by a specific transcription activator. Their screen is based on the use of expression libraries in which the *lacZ* fusion gene is placed under control of yeast regulatory elements. In a pilot experiment, which was carried out to identify genes regulated by the CCAAT-box binding protein, Hap2p (Olesen *et al.*, 1987), 26 fusions regulated by this activator were identified including two ORFs of unknown function from yeast chromosomes III and XI. While this was an early and useful proof of the efficiency of grouping genes of related function based on co-regulation by the same transcriptional activator, the work of DeRisi *et al.* (1997) on Tup1p and Yap1p (see above) indicates that such studies are likely to be pursued at the transcriptional level in the future. Gene fusions, particularly to the coding sequence for Green Fluorescence Protein (GFP; for review, see Cubbitt *et al.*, 1995) will be used to study protein localisation and to monitor gene expression in living cells. An ingenious application of the latter approach is the work of Walmsley *et al.* (1997) in which GFP expression has been placed under the control of the *RAD54* promoter, thus enabling the real-time monitoring of chemical agents or gene mutations that affect the level of DNA damage in the yeast cell.

Proteome analysis

Another route to the elucidation of gene function is the analysis of all yeast proteins synthesised under a given set of conditions, the so-called 'proteome' (Wilkins *et al.*, 1996). The yeast proteome is being defined by two-dimensional gel electrophoreses (Sagliocco *et al.*, 1996) using mass spectrometry to identify the proteins within the spots on the gel (Shevchenko *et al.*, 1996; Fey *et al.*, 1997). By taking full advantage of the complete genome sequence, it should be possible to determine a single protein fragment of unique mass in order to identify a yeast protein. The use of rapid and accurate spectrometric methods should then allow us to monitor changes in the protein maps of yeast strains grown under different physiological conditions, as well as to identify the modification of proteins as a result of the deletion or over-expression of a novel gene. In the long term, this information will enable the identification of all components of protein complexes and thus create a map of all interactions between the yeast proteins, which can be correlated with *in vivo* data obtained from yeast two-hybrid analysis in order to fully understand the working of the yeast cell at the molecular level.

Such a comprehensive two-hybrid approach has already generated a protein linkage map for the *E. coli* bacteriophage T7 genome encoding 55 proteins (Bartel *et al.*, 1996) and a highly selective two-hybrid procedure adapted for exhaustive screens of the yeast genome is currently being developed by Fromont-Racine and co-workers (1997).

Quantitative phenotypic analysis

One reason for taking a quantitative approach to the analysis of gene function is the growing number of genes, found by systematic genome sequencing projects, that have homologues of unknown function in a number of species, but were not previously discovered by classical or "function-first" (Oliver, 1996a) molecular genetics. It is possible that these genes have been missed because quantitative, rather than qualitative, data are required to reveal their phenotypic effect. Another reason to consider a quantitative approach to phenotypic analysis is the high level of redundancy that is apparent in the yeast genome. If a particular gene is a member of a paralogous set of identical (or nearly identical) genes then, provided that they are all regulated in a similar manner and their products are targeted to the same cellular location, the contribution which an individual member of the set makes to the phenotype will, necessarily, be some fraction of the whole.

A conceptual and mathematical framework for the quantitative analysis of gene function is provided by Metabolic Control Analysis (MCA) as pioneered by Kacser and Burns (1973) and by Heinrich and Rapoport (1974). These workers introduced the term flux control coefficient (CJE) to permit the quantitative expression of the proportional contribution which any given enzyme makes to the flux through a metabolic pathway. The sum of all the flux control coefficients (Σ CJE) relevant to a given pathway is 1. Thus, in linear pathways, individual flux control coefficients will normally have values between 0 (no control) and 1 (completely rate-determining enzyme), but in branched pathways values outside this range (i.e. < 0 or > 1) can occur. More detailed reviews of this concept are given by Fell (1992) and Cornish-Bowden (1995). The "top-down" (Brand, 1996) or the related, modular (Rohwer *et al.*, 1996) method to such an analysis has exactly the hierarchical features that are required for a systematic approach to the elucidation of gene function (Oliver, 1996a). This approach aims to divide the metabolic map into a number of large units or modules through, or in, which independent fluxes or metabolite concentrations can be measured. Assignment of a novel gene as having its effect within a particular module then allows that module to be sub-divided into a number of individual units that can be tested, by further flux or metabolite analyses, to determine in which unit the novel gene under examination has its effect. By moving down this hierarchy of analysis, a closer and closer approximation to the function of the novel gene is obtained.

At the highest level of a top-down MCA approach to yeast gene function, we would wish to define the whole cell as the 'system'. This means that we can define the flux control coefficient of a particular gene product by manipulating its cellular concentration and measuring the impact of the change on growth rate, to obtain (in effect) a growth-rate control coefficient. The experimental determination of flux control coefficients (e.g., see Niederberger *et al.*, 1992) demonstrated that they usually have values much closer to 0 than to 1. This means that we need a very sensitive method of measuring the change in growth rate resulting from the specific deletion of a novel yeast gene. Competition experiments between mutant and wild-type yeast have been shown to provide such a sensitive way of measuring small growth rate differences (Danhash *et al.*, 1991).

An attractive approach to competition analysis, called "genetic footprinting", has been developed by Smith *et al.* (1995). In this approach, yeast mutants, generated by Ty1 transposon insertions, were grown in large populations under different selections using serial batch transfers to extend the period of the competition experiments. The relative proportions of the different mutant strains in the population were monitored by PCR using a common primer complementary to the Ty sequence and a series of fluorescently labelled primers complementary to the flanking sequences of the genes whose quantitative phenotype was to be assessed. A sequencing machine equipped with GeneScan™ software was used to perform the analysis of the PCR products. Smith *et al.* (1996) employed the genetic footprinting approach to the quantitative phenotypic analysis of yeast chromosome V. They were able to obtain data for 261 (97%) of the predicted protein-encoding genes which showed, for ca. 60%, a detectable reduction in fitness in one or more of seven different selection protocols; for the remaining genes no phenotypic effect was found. Included in the list of phenotypes expected were novel, or unexpected, effects for a number of known genes. This result suggests that quantitative data obtained by systematic phenotypic analysis can provide important information that will permit, not only the elucidation of the function of novel genes, but will also provide a more integrated view of the cellular role of previously known genes.

In an approach more suited to application in top-down MCA, since measurements of growth-rate differences are made at steady-state in chemostat cultures, Baganz *et al.* (1997a & b) have carried out competition experiments using specific deletion mutants. Control studies showed both that the replacement marker KanMX (Wach *et al.*, 1994) is phenotypically neutral (in contrast to nutritional markers, such as *HIS3* ; Baudin-Baillieu *et al.*, 1996) and that deletion of the *HO* gene by KanMX replacement was also without phenotypic impact (Baganz *et al.*, 1997a). *HO* was chosen because it has no known role, apart from mating-type switching (reviewed by Haber, 1992), and it has been used as the site of insertion of heterologous genes in brewing yeasts without any perceptible effect on the fermentation characteristics of the organism or the quality of the product (Hammond *et al.*, 1994; Yocum, 1986). Whenever possible, a diploid strain, homozygous for the *ho* deletion, is employed - both because *HO* is inactive in diploids and because the mating of independently derived haploid transformants allows any transformation-induced genetic lesions (Danhash *et al.*, 1991) to be nullified through complementation.

In their experiments, Baganz and co-workers compete test strains, bearing KanMX replacements of specific ORFs, with an isogenic standard strain that carries a KanMX replacement of *HO*. This allows the accurate quantitation of the impact of a specific single-gene deletion on growth rate and, moreover, the deletion mutant can subsequently be used for more specific phenotypic analyses. The proportions of mutant and standard-strain cells in the population was measured by quantitative PCR, using either densitometry or the ABI Genescan™ system. Both methods have a rather limited potential for multiplexing although, of course, DNA extracts from cultures may be aliquoted such that multiple PCR reactions may be carried out to compare the proportions of the standard strain and any number of test deletants. However, the best approach for increasing the number of mutants that may be analysed, in a single competition experiment, against the standard strain has been presented by the 'molecular bar coding' technique (Shoemaker *et al.*, 1996) referred to above.

The second type of data required for the MCA approach is the measurement of the change in the relative concentrations of metabolites as the result of the deletion or overexpression of a gene. Teusink *et al.* (1997) proposed an analytical strategy termed FANCY (functional analysis by co-responses in yeast) which is based on the theoretical work by Rohwer *et al.* (1996) and takes advantage of the fact that the number of intermediate metabolites is an order of magnitude lower than the number of genes, and that the functions of about 40% of these genes are already known. Rohwer and co-workers have shown that some enzymes can be grouped into 'monofunctional units' in which any perturbation of the enzymes inside the unit will always produce the same co-responses outside the unit, regardless of which enzyme was affected and irrespective of the magnitude of that perturbation. This property can be exploited for functional analysis by identifying monofunctional units all over the metabolic map of yeast. For example, when two deletion mutants, have the same co-responses, they will affect the same monofunctional unit. If the unit in which one of the genes causes an effect is known, that of the other can be inferred and so the origin of the changes in metabolites located. Such an approach requires a fast and reliable method to measure the concentrations of as many metabolites as possible to produce a 'metabolic snapshot' of each deletion mutant (Teusink et al., 1997). Oliver and co-workers (Oliver, 1997b) are developing a two-stage strategy to obtain such data. In the first phase, deletants of novel genes are grouped with those of known genes by comparing their infra-red spectra under different physiological conditions (Goodacre et al., 1996, Naumann et al., 1996). This information then determines the type of metabolic snapshot to be taken in the second phase, using tandem mass spectrometry (Rashed et al., 1995). The combination of the metabolic profile of a mutant together with its calculated maximum specific growth rate under the same conditions (equivalent to the rate of flux under steady-state conditions) should allow the target site of a novel gene product to be located on the metabolic map of yeast. These measurements are best made at steady-state which can be achieved by using chemostat cultures.

CONCLUSIONS AND PROSPECTS

The new field of Functional Genomics (Hieter & Boguski, 1997) presents yeast researchers, in particular, with new responsibilities and opportunities. The responsibility is to elucidate the function of each and every one of the approximately 4,000 novel protein-encoding genes discovered by the genome sequencing project. The opportunity is to determine how all yeast genes, both those that were discovered by classical ('function first') genetics and those that were revealed by the complete *S. cerevisiae* genome sequence, interact to allow this simple eukaryotic cell to grow, divide, develop, and respond to environmental change. Thus functional genomics should not only provide essential information about the role of novel genes, it should also throw new light on the contributions made by the 'old' genes. If this holistic, or fully integrative, view of the yeast cell can be obtained, one cannot doubt that it will provide an important guide to our understanding of higher eukaryotes such as our own species and our crop plants and farm animals.

ACKNOWLEDGEMENTS

Work on yeast genome analysis in SGO's laboratory has been supported by the European Commission (in both the Yeast Genome Sequencing and EUROFAN Networks), the BBSRC, the Wellcome Trust, and by Pfizer Central Research, Applied Biosystems, Amersham International, and Zeneca Pharmaceuticals. FB is grateful to Pfizer Central Research for the provision of a post-graduate research studentship.

REFERENCES

Amberg, D. C., Botstein, D. and Beasley, E. M. (1995) Precise gene disruption in *Saccharomyces cerevisiae* by double fusion polymerase chain reaction. *Yeast* **11**, 1275-1280.

Bartel, P. L., Roecklein, J. A., SenGupta, D. and Fields, S. (1996) A protein linkage map of *Escherichia coli* bacteriophage T7. *Nature Genetics* **12**, 72-77.

Baganz, F., Hayes, A., Marren, D., Gardner, D. C. J. and Oliver, S.G. (1997a) Suitability of replacement markers for functional analysis studies in *Saccharomyces cerevisiae*. *Yeast* **13**, 1563-1573.

Baganz, F., Hayes, A., Farquhar, R., Gardner, D. C. J. and Oliver, S.G. (1997b) Quantitative analysis of yeast gene function using competition experiments in continuous culture. (In preparation.)

Baudin, A., Ozier-Kalogeropoulos, O., Denouel, A., Lacroute, F. and Cullin, C. (1993) A simple and efficient method for direct gene deletion in *Saccharomyces cerevisiae*. *Nucleic Acids Research* **21,** 3329-3330.

Baudin-Baillieu, A., Guillemet, E., Cullin, C. and Lacroute, F. (1997) Construction of a yeast strain deleted for the *TRP1* promoter and coding region that enhances the efficiency of the polymerase chain reaction-disruption method. *Yeast* **13,** 353-356.

Blattner, F. R., Plunkett, G., Bloch, C. A., Perna, N. T., Burland, V., Riley, M., Collado-Vides, J., Glasner, J. D., Rode, C. K., Mayhew, G. F. *et al.* (1997) The complete genome sequence of *Escherichia coli* K-12 *Science* **277,** 1453-1462.

Bork, P., Ouzounis, C. and Sander, C. (1994) From genome sequences to protein function. *Current Opinion in Structural Biology* **4,** 393-403.

Brand, M. D. (1996) Top-down metabolic control analysis. *Journal of Theoretical Biology* **182,** 351-360.

Burns, N., Grimwade, B., Ross-Macdonald, P. B., Choi, E. Y., Finberg, K., Roeder, G. S. and Snyder, M. (1994) Large-scale analysis of gene expression, protein localization, and gene disruption in *Saccharomyces cerevisiae*. *Genes & Development* **8,** 1087-1105.

Cornish-Bowden, A. (1995) Metabolic control analysis in theory and practice. *Advances in Molecular & Cellular Biology* **11,** 21-64.

Cubbitt, A. B., Heim, R., Adams, S. R., Boyd, A. E., Gross, L. A. and Tsien, R. Y. (1995) Understanding, improving and using green fluorescent proteins. *Trends in Biochemical Sciences* **20,** 448-455.

Danhash, N., Gardner, D. C. J. and Oliver, S. G. (1991) Heritable damage to yeast caused by transformation. *Bio/Technology* **9,** 179-182.

Dang, V. D., Valens, M., Bolotin-Fukuhara, M. and Daignan-Fornier, B. (1994) A genetic screen to isolate genes regulated by the yeast CCAAT-box binding protein Hap2p. *Yeast* **10,** 1273-1283.

Delseny, M., Cooke, R., Comella, P., Wu, H. J., Raynal, M., and Grellet, F. (1997) The *Arabidopsis thaliana* genome project. *Comptes Rendus Hebdomadaires des Seances, Academie des Sciences, Paris* (III) **320,** 589-599.

DeRisi, J. L., Vishwanath, R. I. and Brown, P. O. (1997) Exploring the metabolic and genetic control of gene expression on a genomic scale. *Science* **278,** 680-686.

Dujon, B. (1996) The yeast genome project: What did we learn? *Trends in Genetics* **12,** 263-270.

Fell, D. A. (1992) Metabolic control analysis: a survey of its theoretical and experimental development. *Biochemical Journal* **286,** 313-330.

Fey, S. J., Nawrocki, A., Larsen, M. R., Gorg, A., Roepstorff, P., Skews, G.N.,Williams, R. and Larsen, P. M. (1997) Proteome analysis of *Saccharomyces cerevisiae*: A methodological outline. *Electrophoresis* **18,** 1361-1372.

Foury, F. (1997) Human genetic diseases: A cross-talk between man and yeast *Gene* **195**, 1-10.

Fromont-Racine, M., Rain, J. C. and Legrain, P. (1997) Toward a functional analysis of the yeast genome through exhaustive two-hybrid screens. *Nat. Genet.* **16**, 277-282.

Goffeau, A., Barrell, B. G., Bussey, H., Davis, R. W., Dujon, B., Feldmann, H., Galibert, F., Hoheisel, J. D., Jacq, C., Johnston, M. *et al.* (1996) Life with 6000 genes. *Science* **274**, 546-567.

Goffeau A., Aert, R., Agostine-Carbone, A., Ahmed, A., Aigle, M., Alberghina, L., Albermann, K., Albers, M., Aldea, M., Alexandraki, D. *et al.* (1997) The yeast genome directory. *Nature* **387**, 5-105.

Goodacre, R., Timmins, E. M., Rooney, P. J., Rowland, J. J. and Kell, D. B. (1996) Rapid identification of streptococcus and enterococcus species using diffuse reflectance-absorbency Fourier-transform Infrared-spectroscopy and artificial neural networks. *FEMS Microbiology Letters* **140**, 233-239.

Güldener, U., Heck, S., Fiedler, T., Beinhauer, J. and Hegemann, J. H. (1996) A new efficient gene disruption cassette for repeated use in budding yeast. *Nucleic Acids Research* **24**, 2519-2524.

Haber, J. E. (1992) Mating-type gene switching in *Saccharomyces cerevisiae. Trends in Genetics* **8**, 446-452.

Hammond, J. R. M., Lancashire, W. D., Meaden, P. G., Oliver, S. G. and Smith, N. A. (1994) Stability of genetically modified yeasts in relation to beer of good and consistent quality. MAFF (Ministry of Agriculture, Fisheries and Food; UK) Report 07/63M.

Heinrich, R. and Rapoport, T. (1974) A linear steady-state treatment of enzymatic chains. *European Journal of Biochemistry* **42**, 89-95.

Hieter, P. and Boguski, M. (1997) Functional genomics: It's all how you read it. *Science* **278**, 601-602.

Jia, Y. K., Bécam, A. M. and Herbert, C. J. (1997) The *CIT3* gene of *Saccharomyces cerevisiae* encodes a second mitochondrial isoform of citrate synthase. *Molecular Microbiology* **24**, 53-59.

Jiminez, A. and Davis, J. (1980) Expression of transposable antibiotic resistance elements in *Saccharomyces cerevisiae. Nature* **287**, 869-871.

Kacser, H. and Burns, J. A. (1973) The control of flux. *Symposia of the Society of Experimental Biology* **27**, 65-104.

Kal, A. J. (1997) Transcriptional regulation of genes encoding peroxisomal proteins in *Saccharomyces cerevisiae.* Ph. D. Thesis, University of Amsterdam.

Keogh, R. S., Seoghe, C. and Wolfe, K. H. (1997) Evolution of gene order and chromosome number in *Saccharomyces, Kluyveromyces,* and related fungi. *Yeast* (In press.)

Kunst, F., Ogasawara, N., Moszer, I., Albertini, A. M., Alloni, G., Azevedo,

V., Bertero, M. G., Bessieres, P., Bolotin, A., Borchert, S. *et al.* (1997) The complete genome sequence of the Gram-positive bacterium *Bacillus subtilis. Nature* **390**, 249-256.

Lalo, D., Stettler, S., Mariotte, S., Slonimski, P. P. and Thuriaux, P. (1993) 2 yeast chromosomes are related by a fossil duplication of their centromeric regions. *Comptes Rendus Hebdomadaires des Seances, Academie des Sciences, Paris* (III) **316,** 367-373.

Längle-Rouault, F. and Jacobs, E. (1995) A method for performing precise alterations in the yeast genome using a recyclable selectable marker. *Nucleic Acids Research.* **23,** 3079-3081.

Leong-Morgenthaler, P., Oliver, S. G., Hottinger, H. and Söll, D. (1994) A *Lactobacillus* nifS-like gene suppresses an *Escherichia coli* transaminase B mutation. *Biochimie* **76,** 45-49.

Lockhart, D. J., Dong, H., Byrne, M. C., Follettie, M. T., Gallo, M. V., Chee, M. S., Mittmann, M., Wang, C., Kobayashi, M., Horton, H. and Brown E. L. (1996) Expression monitoring by hybridization to high-density oligonucleotide arrays. *Nature Biotechnology* **14,** 1675-1680.

Metha, P. and Christen, P (1993) Homology of pyridoxal-5'-phosphate-dependent aminotransferases with the *cob*C (cobalamin synthesis), *nif*S (nitrogen fixation), *pab*C (p-amino benzoate synthesis) and *mal*Y (abolishing endogenous induction of the maltose system) gene products. *European Journal of Biochemistry* **211,** 373-376.

Mewes, H.W., Albermann, K., Bähr, M., Frishman, D., Gleissner, A., Hani, J.Heumann, K., Kleine, K., Maierl, A., Oliver, S.G., Pfeiffer F., and Zollner, A. (1997) Overview of the yeast genome. *Nature* **387** (Supp.), 7-65.

Mortimer, R. K. and Johnston, J. R. (1986) Genealogy of principal strains of the yeast genetic stock center. *Genetics* **113,** 35-43.

Mortimer, R. K., Contopoulou, C. R. and King, J. S. (1992) Genetic and physical maps of *Saccharomyces cerevisiae.* Edition 11. *Yeast* **8,** 817-902.

Naumann, D., Schultz, C. and Helm, D. (1996) What can infrared spectroscopy tell us about the structure and composition of intact bacterial cells? In Mantsch, H. H. and Chapmann, D. (Eds) *Infrared spectroscopy of biomolecules.* Wiley-Liss, New York, 279-310.

Niederberger, P., Prasad, R., Miozzari, G. and Kacser, H. (1992) A strategy for increasing an in vivo flux by genetic manipulations - the tryptophan system of yeast. *Biochemical Journal* **287,** 473-479.

Newlon, C. S. (1988) Yeast chromosome replication and segregation. *Microbiolical Reviews* **52,** 568-601.

Oka, A., Sugisaki, H. and Takanami, M. (1981) Nucleotide sequence of the kanamycin resistance transposon *Tn*903. *Journal of Molecular Biology* **147,** 217-226.

Olesen, J., Hahn, S. and Guarente, L. (1987) Yeast HAP2 and HAP3 activators bind to the CYC1 activation site, UAS2, in an independent manner. *Cell* **51,** 953-961.

Oliver, S.G., van der Aart, Q. J. M., Agostoni-Carbone, M. L., Aigle, M., Alberghina, L., Alexandraki, D., Antoine, G., Anwar, R., Ballesta, J. P. G., Benit, P. (1992) The complete DNA sequence of yeast chromosome III. *Nature* **357**, 38-46.

Oliver, S.G., James, C. M., Gent M. E., and Indge, K. J. (1993) The great wall: sequencing the yeast genome. In Heslop-Harrison, J. S. and Flavell, R. B. (Eds), *The Chromosome*. βios Science Publishers, Oxford, 233-245.

Oliver, S. G. (1996a) From DNA sequence to biological function. *Nature* **379**, 597-600.

Oliver, S. G. (1996b) A network approach to the systematic analysis of yeast gene function. *Trends in Genetics* **12**, 241-242.

Oliver, S. G. (1997a) From gene to screen with yeast. *Current Opinion in Genetics & Development* **7**, 405-409.

Oliver, S. G. (1997b) Yeast as a navigational aid in genome analysis. *Microbiology-* **143**, 1483-1487

Olson, M.V. (1991) Genome structure and organization in *Saccharomyces cerevisiae*. In: *The molecular and cellular biology of the yeast Saccharomyces cerevisiae*, J. Broach, J. R., Pringle, and E. W. Jones (Eds), New York: Cold Spring Harbor Lab. Press, pp. 1-39.

Ouzounis, C. and Sander, C. (1993) Homology of the NifS family of proteins to a new class of pyridoxal phosphate-dependent enzymes. *FEBS Letters* **322,** 159-164.

Rashed, M. S., Oznand, P. T., Bucknall, M. P. and Little, D. (1995) Diagnosis of inborn errors of metabolism from bloodspots by acylcarnitines and amino acids profiling using automated tandem mass spectrometry. *Pediatric Research* **38**, 324-331.

Rieger, K.-J., Kaniak, A., Coppee, J.-Y., Aljinovic, G., Baudin-Baillieu, A., Orlowska, G., Gromadka, R., Groudinsky, O., di Rago, J.-P. and Slonimski, P.P. Large-scale phenotypic analysis - the pilot project on yeast chromosome III. *Yeast* (In Press).

Rohwer, J. M., Schuster, S. and Westerhoff, H. V. (1996) How to recognize monofunctional units in a metabolic system. *Journal of Theoretical Biology* **179,** 213-228.

Ross-Macdonald, P., Erdman, S., Agarwal, S., Burns, N., Sheehan, A., Malczynski, M., Roeder, G. S. and Snyder, M. (1997) Large scale functional analysis of the *Saccharomyces cerevisisae* genome. *Yeast* **13**, S10.

Rothstein, R. J. (1983) One-step gene disruption in yeast. *Methods in Enzymology* **101,** 3694-3698.

Sagliocco, F., Guillemot, J., Monribot, C., Capdevielle M., Perrot, M., Ferran, E., Ferrara, P. and Boucherie, H. (1996) Identification of proteins of the yeast protein map using genetically manipulated strains and peptide-mass fingerprinting. *Yeast* **12,** 1519-1534.

Schena, M., Shalon, D., Heller, R., Chai, A., Brown, P. O. and Davis, R. W. (1996) Parallel human genome analysis-microarray-based expression monitoring of 1000 genes. *Proceedings of the National Academy of Science, USA* **93**, 10614-10619.

Shevchenko, A., Jensen, O. N., Podtelejnikov A. V., Sagliocco, F., Wilm, M., Vorm, O., Mortensen, P., Shevchenko, A., Boucherie, H. and Mann, M. (1996) Linking genome and proteome by mass spectrometry - large-scale identification of yeast proteins from 2-dimensional gels. *Proceedings of the National Academy of Science, USA* **93,** 14440-14445.

Shoemaker, D. D., Lashkari, D. A., Morris, D., Mittmann, M. and Davis, R. W. (1996) Quantitative phenotypic analysis of yeast deletion mutants using a highly parallel molecular bar-coding strategy. *Nature Genetics* **14,** 450-456.

Smith, V., Botstein, D. and Brown, P. O. (1995) Genetic footprinting - a genomic strategy for determining a gene's function given its sequence. *Proceedings of the National Academy of Science, USA* **92,** 6479-6483.

Smith, V., Chou, K. N., Lashkari, D., Botstein, D. and Brown, P. O. (1996) Functional analysis of the genes of yeast chromosome V by genetic footprinting. *Science* **274,** 2069-2074.

Steiner, S. and Philippsen, P. (1994) Sequence and promoter analysis of the highly expressed *TEF* gene of the filamentous fungus *Ashbya gossypii. Genetics* **119,** 249-260.

Sun, D. and Setlow, P. J. (1993) Cloning and nucleotide sequence of the *Bacillus subtilis nad*B gene and a *nif*S-like gene both of which are essential for NAD biosynthesis. *Journal of Bacteriology* **175,** 1423-1432.

Teusink, B., Baganz, F., Westerhoff, H. V. and Oliver, S. G. (1998) Metabolic control analysis as a tool in the elucidation of the function of novel genes. In: *Methods in Microbiology* **26,** 297-336.

Velculescu, V. E., Zhang, L., Vogelstein, B. and Kinzler, K. W. (1995) Serial analysis of gene expression. *Science* **38,** 484-487.

Velculescu, V. E., Zhang, L., Vogelstein, J., Basrai, M. A., Bassett D. E. Jr, Hieter, P., Vogelstein, B. and Kinzler, K. W. (1996) Characterization of the yeast transcriptome. *Cell* **88,** 243-251.

Wach, A., Brachat, A., Pohlmann, R. and Philippsen, P. (1994) New heterologous modules for classical or PCR-based gene disruptions in *Saccharomyces cerevisiae. Yeast* **10,** 1793-1808.

Wach, A. (1996) PCR-synthesis of marker cassettes with long flanking homology regions for gene disruptions in *S. cerevisiae. Yeast* **12,** 259-265.

Wach, A., Brachat, A., Alberti-Segui, C., Rebischung, C. and Philippsen, P. (1997) Heterologous *HIS3* marker and GFP reporter modules for PCR-targeting in *Saccharomyces cerevisiae. Yeast* **13,** 1065-1075.

Walmsley, R. M., Billinton, N. and Heyer, W.-D. (1997) Green fluorescent protein as a reporter for the DNA damage-induced gene *RAD54* in *Saccharomyces* cerevisiae. *Yeast* (In press.)

Waterson, R. and Sulston, J. (1995) The genome of *Caenorhabditis elegans. Proc. Nat. Acad. Sci. USA* **92,** 10836-10840.

Wilkins, M. R., Pasquali, C., Appel, R. D., Ou, K., Golaz, O., Sanchez, J. C., Yan, J. X., Gooley, A. A., Hughes, G., Humphrey-Smith, I. *et al.* (1996) From proteins to proteomes - large-scale protein identification by 2-dimensional electrophoreses and amino-acid-analysis. *Bio/Technology* **14,** 61-65.

Wolfe, K. H. and Shields, D. C. (1997) Molecular evidence for an ancient duplication of the entire yeast genome. *Nature* **387,** 708-713.

Yocum, R. (1986) Genetic engineering of industrial yeasts. *Proceedings Bio Expo* **86,** 17.

Zheng, L., White, R. H., Cash, V. L., Jack, R. F., and Dean, D. R. (1993) Cysteine desulfurase activity indicates a role for NifS I metallocluster biosynthesis. *Proceedings of the National Academy of Science, USA* **90,** 2754-2758.

GENE FUNCTIONALITY

FUNCTIONAL ANALYSIS OF HUMAN GENES

C.T. CASKEY, Q. LIU, M.L. METZKER, D. GERHOLD AND C.P. AUSTIN

Merck & Co., Inc., Human Genetics Laboratory, P.O. Box 4, West Point, Pennsylvania 19486, USA

ABSTRACT

Merck committed to the establishment of an open Expressed Sequence Tag (EST) database four years ago and has been the major contributor to the unrestricted database over the past three years. The daily searching of that database is now routine. This data base has altered the paradigm of candidate disease gene identifications, expansion of gene families, and gene associations between model genomes and man. We have utilized this database for our studies of full length cDNAs with relevance to drug development targets. Toward that objective, a simple and automated method of full length cDNA was developed and is routinely implemented in our drug discovery programs. We utilize routinely fluorescent dyes with signal uniformity and greater sequence reads toward the objective of functional genomics. This latter challenge has been approached in several ways. A chip expression database for the "genes of interest" has been established with both Synteni and Affymetrix. The challenge of <u>sequence to function</u> is considerable. Toward this objective, Merck formed the Merck Genome Research Institute to stimulate development of new technology which will be available to all investigators. The first of these major commitments have been made in the cDNA, informatics, yeast, *Drosophila*, and mouse areas. It is clear that genome science is making major contributions to drug development at Merck. It should be remembered, however, that > 90% of discoveries that lead to drug development are made in the academic research arena, while drug development is almost exclusively conducted in the industrial setting. The functional study of our fixed number of genes needs to be freely accessible; i.e. genes are research tools for drug development. We continue our policy of providing unincumbered gene information and research tools toward the objective of acceleration sequence → function. The number of chemical entities which are potential drugs is extremely high given combinatorial chemistry and thus efficacy, toxicity, and unwanted actions will vary considerably for a single gene target. An exception to this rule is gene therapy and protein therapy where the gene (possibly) with some slight variations serves as the therapeutic product. Even in this case, presentation, expression, targeting, and formulation hold keys to the product success. The access or lack of access to genes as research tools will accelerate or complicate drug discovery and the impact of genome science on health.

INTRODUCTION

The human genome initiative has a stated objective of completing the 3.0×10^9 base pair human sequence and selected model systems. It is the knowledge of the gene function(s) in health and disease which will alter our experimental approaches to molecular, cellular, and *in vivo* biology. This new scientific wave is forming and its crest is anticipated to change our acceleration toward gene understanding. In the following sections the applications of Genome Science within Merck will be discussed.

THE MERCK GENE INDEX

The Merck Gene Index, having been in operation for approximately three years, has contributed over 480,000 EST entries into the public database. This continuing effort has been carried out by the IMAGE consortium, with the sequencing activities focused at Washington University Genome Center. There were several objectives to this project which have in large part been achieved. These objectives include: 1) A focus on human gene sequences and 2) an unencumbered accessible gene database to accelerate disease gene discovery and gene family expansions. This Index is now a routine research tool for genome scientists and has led to rapid and facile discovery of a plethora of disease genes. The value of the Index has been enhanced by improved public informatics and continued chromosomal mapping of many (~16,000) ESTs. Thus in position cloning, it is routine to focus first on the likely gene suspects by a database search of ESTs mapped to the disease locus and, secondly, to match ESTs of unknown function to the genomic sequence obtained from the disease locus. (Aaronson *et al.*, 1996; Hillier *et al.*, 1996; Schuler *et al.*, 1996) Furthermore, one now routinely expands a gene family by database searching of ESTs, regardless of map position, using a variety of algorithms. Present focus is on use of the sequences and their full length cDNAs and also their model system analogues to estimate the function of the genes whose function or pathways are not elucidated. While the task of the EST database is not complete, Merck is now initiating overlapping initiatives in the area of Functional Genomics. The activity is sponsored by the Merck Genome Research Institute, a non-profit Institute dedicated to the development of methods to facilitate our conversion of gene sequence to function. Initiatives are now funded in the cDNA, informatics, model system comparisons, yeast 2 hybrid, and genome-wide mouse knockout technology.

MINING THE DATABASE

Cloning and characterization of receptor genes have been of particular interest to pharmaceutical companies. Historically, receptors genes have been isolated either by expression cloning, oligonucleotides (based on amino acid sequence) hybridization, (Dixon *et al.*, 1986; Kubo *et al.*, 1986) low stringency cross-hybridization, and degenerate PCR guided by cloned receptor sequences. While these approaches are still employed to clone receptors for specific ligands, the recent explosion in the number of ESTs, by both public and private efforts, provides a rich source for the identification and cloning of novel receptor genes in a more efficient, high throughput manner.
Sequence analysis of cloned receptor genes has led to the classification of genes into families based on sequence homology. Each family has its distinct sequence or structural

motifs which are highly conserved among members of the family. These motifs are often essential for gene functions, as shown by extensive analysis using site-directed mutagenesis. Furthermore, these motifs are often so unique as to predict functions of novel sequences. Therefore, searching for the presence of family-specific motifs among the massive number of EST sequences can quickly lead to the discovery of ESTs that are likely to encode novel members of the gene family (Venter, 1993). Furthermore, some genes which were long thought to be unique turned out to have closely related members, as illustrated by the recent cloning of a new uncoupling protein gene (UCP2) highly homologous to the uncoupling protein (UCP1) of brown fat mitochondria (Fleury et al., 1997, Gimeno et al., 1997).

Searching sequence databases for homologous sequences is most commonly carried out by the Fasta or Blast program (Pearson and Lipman, 1988; Altschul et al., 1990). Queries using amino acid sequences are much more sensitive to weak homologies. Automated searches were utilized for the gene families of interest. For each family, all the protein sequences of the known members of that family were collected and used to search the daily EST additions to the EST database, after an initial search against the entire existing database was completed. EST hits above selected thresholds were selected and used to search against sequences of all the known genes to confirm that they are significantly homologous to, but different from, the query sequences. These novel EST sequences were then collected and used to isolate their full-length genes.

Since the majority of ESTs in the database were derived from short, oligo-dT primed cDNA clones, ESTs often provide partial cDNA sequences only (Soares et al., 1994). A high throughput method to extend partial cDNA sequences rapidly is needed to isolate full-length clones for the large number of ESTs found. The traditional process of isolating cDNA clones by hybridization has changed little since its inception, two decades ago. The method is time-consuming and labor intensive, thus unable to handle large numbers of genes. More recently, developed techniques such as 5' and 3' race, rely on PCR to amplify the target sequence from total cellular mRNA. Such PCR-based methods offer much increased speed and decreased labor. Unfortunately, the methods are limited in amplifying full-length gene sequences due to the competitive amplification of short PCR products. Recently, pooling strategies were described to achieve rapid isolation of cDNA clones (Munroe et al., 1995; Alpheny, 1997). These methods, however, require a considerable amount of effort in pooling and superpooling the cDNA libraries and hybridization to identify the clones.

We have developed a PCR-based process for rapid extension of EST sequences for genes expressed at low levels. This process, designated Reduced Complexity cDNA Analysis (RCCA), relies on PCR amplification in cDNA pools with reduced complexity. Primary cDNA libraries were amplified at approximately 10,000 clones/plate on solid medium. The clones on each plate were collected and stored in 96-well plates. A library comprising one million primary clones, for example, could be amplified on 96 plates and stored in a single 96-well plate. EST-positive wells containing cDNA clones that correspond to an EST sequence were identified by PCR amplification using a pair of EST-specific primers. For genes expressed at low levels, each positive well is unlikely to contain more than one cDNA clone corresponding to a given EST sequence. cDNA fragments flanking the EST sequences were isolated from those positive wells by PCR amplification using vector and insert primers. For libraries constructed in circular plasmid vectors, cDNA clones within positive pools can be recovered by inverse PCR using a pair of EST-specific primers that lie back-to-back on the EST sequence followed by self-ligation and transformation of the PCR product. The cDNA fragments derived from vector-insert primer PCR or from inverse PCR, or both, are sequenced and assembled with the original EST sequence. The steps are repeated by "walking" with primers designed from new end sequences if necessary. In comparison with previous PCR-based methods, this process offers several advantages: 1) Pooling from primary libraries instead of amplified libraries gives better coverage; 2) for low to rare abundance genes, each pool is unlikely to contain more than one target cDNA clone, making it possible to extend the EST sequence from different cDNA clones simultaneously. Analysis of different clones allows the identification of the largest clone as well as alternatively or incompletely spliced forms. Furthermore, sequencing of fragments from different cDNA clones often results in rapid sequence assembly because the clones end at different places; 3) The full-length coding sequence can be amplified by high-fidelity PCR from specific pools after the sequencing analysis is complete. Alternatively, the original cDNA clone can be retrieved from specific pools by standard hybridization method if necessary. Since the RCCA process narrows down to ~10,000 clones with the complete gene sequence determined simultaneously, the method remains quite useful, even in cases where hybridization has to be performed eventually.

Using this strategy as well as other standard cDNA isolation techniques, we have isolated and sequenced the complete coding regions for a large number of genes belonging to the various gene families. These genes are being characterized for their expression pattern at the tissue level as well as at the cellular level. Function analyses of these genes are in progress and the understanding of new gene family members is highly likely to identify novel pathways and, consequently, therapeutic targets.

THE SEQUENCING ENGINE

The ultimate goal of the human genome project is gene discovery that could significantly increase our understanding of human biology and pathogenesis. The complete sequencing of the three billion base-pairs of the human genome is expected to be finished by the year 2005. As of June 1997, approximately 1.5 percent of the genome exists in the public databases, and it is anticipated that that number will grow to 7 percent by June 1998. The scaleability of the current automated technology to generate 100 percent of the sequencing data over the remaining seven years is achievable, and the progress of high-throughput sequencing can be monitored easily by public database submissions. Accurate DNA sequence analysis and annotation of the entire human genome, however, poses formidable challenges for current informatic analyses. In fact, the existing bioinformatic tools are inadequate in extracting gene function information from "finished" DNA sequencing data. The completion of the sequencing phase of the human genome project should therefore mark the beginning of large-scale functional analyses of the human genome. The application of this genome information towards drug and gene therapeutic intervention will carry on well into the twenty-first century. This section will attempt to review, albeit not comprehensively, the current and future directions of sequencing technologies of (ESTs) and full-length cDNAs in the public forum in promoting gene discovery.

The utility of ESTs for gene discovery is almost universally accepted, yet, conservatively, only 50 percent of the genes may be represented by ESTs in the public domain. Cluster analyses of human ESTs have reduced the data set to approximately 43,000 to 49,000 unique transcripts (Aaronson *et al.*, 1996; Hillier *et al.*, 1996; Schuler *et al.*, 1996).

Cross-species comparisons with mouse, *Drosophila*, *Caenorhabditis elegans*, yeast, *Arabidopsis*, and microbial genomes have also provided invaluable insight of gene function for human ESTs and full-length genes. In 1996, the Howard Hughes Medical Institute (HHMI) initiated a 5' end mouse EST project run by GSC to expand these studies. In cooperation with the IMAGE consortium, over 170,000 ESTs have been deposited into dbEST.

The National Cancer Institute (NCI) has created the Cancer Genome Anatomy Project (CGAP) to focus on cancer genes for prostate, breast, colon, lung and ovaries. In addition to EST sequencing from normalized tissue-specific libraries, laser capture microdissection (Emmert-Buck *et al.*, 1996) has been used to isolate normal and tumorous "homogeneous" cell populations, initially from human prostate tumor. As few as 5-10,000 cells are then used to construct libraries and ESTs are sequenced and deposited into dbEST by the GSC-IMAGE consortium. Preliminary results from several thousand ESTs generated by NCI-CGAP, show approximately 25% of the ESTs have no significant database matches. These results are encouraging and highlight the need for more diverse tissue-specific libraries in order to obtain coverage of every gene in the human genome. These initiatives illustrate the continued improvement in the mammalian gene databases and the impact of new technology to achieve greater depth in the database.

The public momentum of EST sequencing is shifting to full-length insert or full-length cDNA sequencing. Both NCI-CGAP and The Institute for Genomic Research (TIGR) have recently announced plans to sequence 10,000 full-length genes over the next several years. To coordinate these and the efforts of many genome sequencing centers, the Department of Energy recently held the *Workshop on Complete cDNA Sequencing*. Unlike ESTs, however, current technologies for rapid and efficient sequencing of full-length cDNAs are still in the piloting stages for library construction, DNA sequencing, and full-length validation. For example, several library strategies have been described to capture the full-length cDNA, such as the oligo-capping (Maruyama and Sugano, 1994), CAPture (Ederly *et al.*, 1995), CapFinder (CLONTECHniques, 1996), and CAP trapper (Carninci *et al.*, 1996) methods. While these methods focus on obtaining the full-length cDNA clone, the enhancement of these methods over conventional or sizing strategies remains obscure. With this in mind, many research groups are moving ahead to rapidly sequence full-length inserts compared to full-length cDNAs.

The same issues of speed, efficiency, scaleability, accuracy, and costs that most high-throughput genome sequencing centers faced several years ago will most likely determine the technology used to sequence thousands of full-length inserts/cDNAs, referred to here as full-length clones. A number of strategies are currently being evaluated including primer walking, transposon-based, and concatenation of multiple full-length clones (Andersson *et al.*, 1996). The former two are very efficient methodologies, but the issues of speed, scaleability, and costs may seriously impede the use of these protocols in high throughput sequencing of full-length clones. The concatenated cDNA strategy, on the other hand, relies on conventional "shotgun" sequencing technologies that the majority of genome sequencing centers have successfully used in their high-throughput operations.

Size validation of the full-length gene poses the greatest challenge for high-throughput analyses of multiple full-length cDNA because of the difficulty of scaling numerous Northern analyses. Moreover, repeated sampling of the library for the remaining 5' end of the gene will inherently drive the efficiency of the effort below acceptable levels. Although it is by far not the best solution, acceptance of 90, 80, or even 70 percent coverage of the full-length gene as a first pass for high-throughput analyses may be sufficient in facilitating gene discovery efforts beyond EST sequencing. In some cases genes are extensively alternatively spliced leading to significant functional differences. In such cases complete genomic sequence facilitates knowledge of spectrum and functionality.

It would be short-sighted to focus exclusively on the full-length gene sequences, since each has a regulation program which is special to its function and *in vivo* integration in the mouse or human. We have elected to focus our activities on a selected number of genes within drug development candidates to assist in the target validation and characterization. A genome wise matching of ESTs to regulatory domains would be highly recommended as a limited sequencing project, designed to characterize a functional element.

FUNCTIONAL GENOMICS

Given the large size of the human and model organism genomes being sequenced, the challenge of determining the function of all the genes in these organisms will be considerable. Technologies to assign function will thus need to be simple, robust and high-throughput. Cell and organismic biologists will need to adapt to the scale and challenges of genome science, as the molecular biologist did before them. Approaches being used to assign gene function include: 1) Bioinformatic associations within and between genomes; 2) determination of gene expression pattern by tissue- and gene-chip-based technologies; 3) germ line modification of gene expression by mouse knock-out, knock-in and transgenic strategies, and 4) somatic modification of gene expression by conditional knock-out in mice, antisense/ribozyme methods and viral gene transfer. Each has some utility in the validation of drug targets and we have incorporated all into our Genome Program, but none satisfies the requirements of simplicity, robustness and throughput.

The study of a gene's expression pattern in health and disease often gives clues as to its function. Two complementary methods to establish these patterns are *in situ* hybridization and DNA chip technologies. *In situ* hybridization allows the localization of expression of a gene to a specific cell type in a specific tissue, and current non-isotopic methods are sensitive down to a few copies per cell. Changes in the quantity or distribution of gene expression in a disease state, or co-localization of transcripts or antibodies, can be extremely useful information in the determination of a novel gene's function. Such tissue-based methods are necessarily labor-intensive and low-throughput compared to the DNA chips (made by Affymetrix, Inc., Synteni, Inc., and others), which allow the monitoring of expression levels of several thousand genes simultaneously down to a single copy per cell (Maier *et al.*, 1997). Though no information on cell type specificity can be obtained, gene expression microaways or "DNA chips" are perhaps the first high-throughput tools for functional genomics. By arraying thousands of genes on a glass chip and hybridizing with expressed genes from tissues of interest, one can determine which genes are expressed and regulated in, for example, normal *versus* diseased tissues.

Merck's collaboration with Synteni has allowed the presentation of ten thousand cDNAs of 300-3,000 base pairs each, on each gene chip. Pairs of probes are made from two tissues, e.g. normal and diseased brain, and labeled with two different colored fluorescent dyes. A probe pair is co-hybridized to a single chip. The ratio of colors hybridizing to each spot registers which genes are regulated differentially in the diseased tissue.

Affymetrix represents each gene on a chip as 40 or more 25mer oligonucleotides. Fluorescent probes are made similarly from tissues of interest, hybridized to individual chips, and expression of each gene is determined by summing the 40 or more data points for each gene. Use of Affymetrix 65,000 oligonucleotides per chip format is also finding applications in detection of mutations and polymorphisms in genomic DNA.

Expression chip data at Merck has been used to: 1) profile expression of novel genes in 40 human tissues; 2) evaluate metabolism of drugs by rodent and human livers, and, 3) carry out gene-to-function drug target validation by identifying genes induced in diseased tissues. The latter studies provide a fascinating glimpse of the "molecular mechanics" of a cell's response to disease or drug treatment. With further scaling of these technologies, it may soon be possible to assay the expression of all known human genes simultaneously.

Our experience has lead Merck scientists to the conclusion that functional analytic methods are inadequate to the challenge of determining function. In an effort to improve the tools and approaches to functional genetics, a non-profit institute, The Merck Genome Research Institute, was founded to fund such research in the academic community. An example of such an initiative is the Mouse Omnigene Project conducted by Lexicon, a mouse Biotech Corp. located in The Woodlands, TX, USA. Strategy and progress on this genome wide knockout project will be given. The technology spectrum of other funded projects is wide and will be listed.

CENTRAL NERVOUS SYSTEM DISEASE RESEARCH

Discoveries in human genetics have had a disproportionate impact on nervous system disease research. This may be because many central nervous system (CNS) diseases have considerable genetic contributions to etiology, and because the study of CNS diseases with conventional physiological and pharmacological methods has been difficult due to the inaccessibility of the brain to experimental manipulation.

The contribution of genetics to understanding the neurodegenerative diseases are a case in point. These are late-onset disorders characterized by cell death of selected neuronal populations in the CNS, and have few equivalents in other organ systems.

Though their pathology has been well known for some time, the etiology of these diseases remained completely unknown until genes associated with them were identified. Genes for familial Alzheimer's disease (Sherrington *et al.*, 1995), Parkinson's disease (Polymeropoulos *et al.*, 1997), Huntington's disease (Huntington's Disease Collaborative Reasearch Group, 1993), familial amotrophic lateral sclerosis (ALS) (Rosen *et al.*, 1993), and a variety of spinocerebellar degeneration syndromes (e.g., Banfi *et al.*, 1994) have been isolated in the last five years, and each discovery has afforded fundamental insights into disease mechanism.

The trinucleotide repeat expansion diseases are an excellent example of the impact of genomics efforts on the understanding of human disease and identification of drug targets (Gusella and MacDonald, 1996). Benign di- and tri-nucleotide repeats are common throughout the human genome, and serve as the most useful genetic markers. In proximity to certain genes, however, they can cause diseases, which, for unknown reasons, virtually all affect the nervous system. These repeats may be in the 5' UTR (CGG repeats in Fragile X syndrome), coding regions (CAG repeat diseases; see below), introns (GAA repeats in Freidriech's ataxia) or 3' UTR (CTG repeats in myotonic dystrophy). The CAG repeat diseases, the most common of which is Huntington's disease, are characterized by CAG repeat exansions in the coding regions of the responsible disease gene, and have been found in six late-onset neurodegenerative diseases characterized by selective cell death. These CAG repeat expansions code for an expanded tract of polyglutamine within the disease protein, which is thought to lead to selective cell death via an unclear mechanism. Because these diseases are dominantly inherited, and the disease proteins are expressed widely, these proteins are thought to interact with other proteins to cause toxicity. Several proteins that interact with polyglutamine-containing disease proteins have been isolated (e.g. Kalchman *et al.*, 1997); though their involvement in disease pathogenesis is unclear, the identification of the genes, their pathogenic mutations, and interacting proteins have provided multiple targets for possible drug development for these previously idiopathic diseases.

Other neurodegenerative diseases are also beginning to yield to genetic approaches. In the last five years several Alzheimer's disease risk genes have been isolated, including the amyloid precursor protein gene, presenilin-1, presenilin-2, and apoE (Cruts *et al.*, 1996). The gene for familial ALS was cloned in 1993 and is a Cu/Zn superoxide dismutase (Rosen *et al.*, 1993), consistent with prior data implicating free radical toxicity in the pathogenesis of ALS. Most recently, mutations in the α-synuclein gene, which codes for a synaptic vesicle protein also identified as a component of Alzheimer amyloid plaques, have been identified in Parkinson's disease patients (Polymeropoulos et al, 1996). Importantly, however, these genes have been isolated from families with rare familial forms of the diseases, and their relevance to the more common sporadic forms is unclear.

The major neuropsychiatric disorders, particularly schizophrenia and bipolar illness, have been the object of intensive positional cloning efforts, but none has been successful to date (Pulver et al, 1997; Risch and Botstein, 1996). This probably stems from the absence of a clear familial autosomal dominant form of the disease (as has been present for the neurodegenerative diseases), difficulties of accurate diagnosis and data collection in these populations, and the multigenic nature of these disorders. It is likely that within the next five years firm chromosomal linkages will be established in these diseases, and perhaps even the first susceptibility genes. The progress in development of the human gene map (Schuler *et al.*, 1996), and the ongoing public and private sequencing efforts (Hillier *et al.*, 1996), should accelerate this process.

Genomic technologies have brought about rapid and fundamental advances in understanding CNS diseases which had resisted all previous attempts at elucidation via conventional non-genetic methods. Some of the susceptibility genes identified may be drug targets themselves, but more likely they will indicate a pathway or general mechanism that can be manipulated for therapeutic purposes. Commercial applications of these discoveries will come not only in improved diagnosis and prognosis, but also in the development of new therapies for increasingly prevalent diseases in the world's aging population.

ETHICAL AND LEGAL ISSUES

A critical issue in the commercial application of genomics discoveries is the proper assignment and utilization of intellectual property (for review see Caskey and Tribble, 1997). Within the past year, the US Patent and Trademark Office (USPTO) has made two announcements which will affect the patentability of genomic inventions and business decisions relating to the use of genomic sequences. First was a decision to allow patenting of partial gene sequences such as expressed sequence tags (ESTs) based on their utility as probes to identify the specific DNA sequences, though this decision may be reversed because of apparent conflict with established patent law (which states that intermediates used to make chemical end-products for which there is no known utility are not allowed) and objections from Harold Varmus (head of the US national Institutes of Health) and others. The Commissioner of patents has stated that there may be instances when broad EST claims are appropriate, and that such claims might preclude future patenting of the full length gene. It is this uncertainty that causes concern to both the scientific and commercial communities involved in cloning of human disease genes. The second major development was that, in response to a large number of patent applications containing thousands of ESTs each, the USPTO enacted a policy of allowing ten sequences (ten distinct inventions) in a single patent application, reversing prior policy that viewed nucleotide sequences encoding different proteins as distinct compounds. This has reduced the cost of acquiring any given EST patent and increased the likelihood that such patents will be widespread.

In contrast to the USPTO, European patent offices have generally required more demonstrated function for patents to be allowed. In an effort to harmonize national rules on patenting of genetic material across Europe, the European Parliament recently approved a controversial Biotechnology Directive. The new directive specifically requires that the industrial application, or utility, of a sequence or a partial sequence of a human gene must be disclosed in the patent application. The standardization of patentability criteria across Europe will allow Biotechnology companies to protect the genomic inventions which will be developed into products, space and, thus, will likely result in an expansion of the European biotechnology industry.

An area that has received little attention from both the scientific and legal communities is the patenting of entire genomes. Genome researchers are currently sequencing numerous microorganisms resulting in entire sequences without much, if any, initial information about the individual genes. In at least one instance (the hepatitis C virus), an entire genome has been patented, with the effect that many laboratories have had to discontinue work on therapeutivs or vaccines. Indeed, if patents continue to be allowed for entire genomes, it could have a chilling effect on the development of new anti-infectives and vaccines.

Scientific discovery has often conflicted with prevailing world views, but genomic technologies have created a new level of ethical and philosophical tension because of their ability not only to explain, but also to change, the natural world in direct and permanent ways. Since the start of the recombinant DNA era there have been attempts to stop all such research, or prohibit the use of genetically altered organisms. These attempts have usually been unsuccessful, because the argument has been made that new strains of organisms created by molecular biology techniques are not fundamentally different from new strains created by selective breeding, which is well established and non-controversial. The recent cloning of a sheep from a single cell, though not genetically manipulated, illustrates the power of modern molecular and cell-based techniques to both create novel organisms and to stir the public's imagination and fears. The advent of genetic testing for untreatable late-onset disorders (e.g., Huntington's disease) has raised questions as to when these tests are appropriate given the lack of therapeutic options. Testing for mutations in susceptibility genes for common diseases (e.g., BRCA 1 and 2) have raised other questions because the likelihood of disease given a positive test is not known, and therapeutic options are not clear. There are fears that employment, or health and/or life insurance will be lost if the results of genetic testing are known; a recent survey (Lapham et al., 1996) indicated a high level of perceived discrimination among those with genetic diseases. A presidential National Bioethics Advisory Commission, headed by Harold Shapiro of Princeton University, has recently been appointed to make recommendations on what government guidelines are needed to prevent abuses of genomic technologies, and the several state legislatures are considering bills outlawing genetic discrimination.

SUMMARY

The human genome initiative is now moving rapidly into the functional and applications era. It is this phase which will give the public benefit. Given the limited number of genes determining normal and disease function it is our opinion that genes should be considered research tools with wide access for study. Since many therapeutic approaches can address a disease it is in the detail of a disease solution that patent protection for the investment can be achieved. Finally, we should be reminded each of us is genetically unique and thus differ in our disease sequenceability. In anticipation of therapeutic interventions of those risks, we should assure a public that genetic testing will be held private, can not be used to discriminate, and will be recommended/used when the patient will benefit from precise diagnosis and/or therapeutic intervention.

Finally, it should be remembered that human research is needed to achieve therapy objectives. Genetic testing for research purposes will accelerate that objective and when conducted has the obligation of informed consent and privacy.

REFERENCES

Aaronson, J.S., Eckman, B., Blevins, R.A., Borkowski, J.A., Myerson, J., Imran, S., and Elliston, K.O. (1996) Towards the Development of a Gene Index to the Human Genome: An Assessment of the Nature of High-throughput EST Sequencing Data. *Genome Research* 6: 829-845.

Alphey, L. (1997). PCR-based method for isolation of full-length clones and splice variants from cDNA libraries. *Biotechniques* 22, 481-4, 486.

Altschul, S.F., Gish, W., Miller, W., Myers, E.W. and Lipman, D.J. (1990). Basic local alignment search tool. *Journal of Molecular Biology* 215, 403-10.

Andersson, B., Lu, J., Shen, Y., Wentland, M.A., and Gibbs, R.A.. (1997) Simultaneous Shotgun Sequencing of Multiple cDNA Clones. *DNA Sequence* 7: 63-70.

Banfi, S., Servadio, A., Chung, M., *et al.* (1994). Identification and characterization of the gene causing type 1 spinocerebellar ataxia. *Nature Genetics* 7:513-519.

Caskey, C.T. and Tribble, J.L. (1997). Who owns the code? *Science and Public Affairs, Autumn*, 52-55.

CLONTECHniques Technical Bulletin (1996) January pp. 2-4

Cruts, M., Hendriks, M. and Van Broeckhoven, C. (1996). The presenilin genes: A new gene family involved in Alzheimer's disease pathology. *Human Molecular Genetics* 5:1449-1455

Dixon, R.A., Kobilka, B.K., Strader, D.J., Benovic, J.L., Dohlman, H.G,. Frielle, T., Bolanowski, M.A., Bennett, C.D., Rands, E., Diehl, R.E., *et al.* (1986). Cloning of the gene and cDNA for mammalian beta-adrenergic receptor and homology with rhodopsin. *Nature* 321, 75-9.

Ederly, I,. Chu, L.L., Sonenberg, N., and Pelletier, J.. (1995) An efficient strategy to isolate full-length cDNAs based on an mRNA cap retention procedure (CAPture). *Molecular and Cellular Biology* 15: 3363-3371

Emmert-Buck, M.R., Bonner, R.F., Smith, P.D., Chuaqui, R.F., Zhuang, Z., Goldstein, S.R., Weiss, R.A., and Liotta, L.A.. (1996) Laser Capture Microdissection. *Science* 274: 998-1001.

Fleury, C., Neverova, M., Collins, S., Raimbault, S., Champigny, O., Levi-Meyrueis, C., Bouillaud, F., Seldin, M.F., Surwit, R.S., Ricquier, D. and Warden, C.H.. (1997). Uncoupling protein-2: a novel gene linked to obesity and hyperinsulinemia [see comments]. *Nature Genetics* 15, 269-72.

Gimeno, R.E., Dembski, M,. Weng, X., Deng, N., Shyjan, A.W., Gimeno, C.J., Iris, F., Ellis, S.J., Woolf, E.A. and Tartaglia, L.A.. (1997). Cloning and characterization of an uncoupling protein homolog: a potential molecular mediator of human thermogenesis. *Diabetes* 46, 900-6.

Gusella, J. and MacDonald, M. (1996). Trinucleotide Instability: a repeating theme in human inherited disorders. *Annual Review of Medicine* 47:201.

Hillier, L., Lennon, G., Becker, M., Bonaldo, M.F., Chiapelli, B., Chissoe, S., Dietrich, N., DuBuque, T., Favello, A., Gish, W., Hawkins, M., Hultman, M., Kucaba, T., Lacy, M., Le, M., Le, N., Mardis, E., Moore, B., Morris, M., Parsons, J., Prange, C., Rifkin, L., Rohlfing, T., Schellenberg, K., Soares, M.B., Tan, F., Thierry-Meg, J., Trevaskis, E., Underwood, K., Wohldman, P., Waterston, R., Wilson, R., and Marra, M.. (1996) Generation and Analysis of 280,000 Human Expressed Sequence Tags. *Genome Research* 6: 807-828.

The Huntington's Disease Collaborative Research Group (1993). A novel gene containing a trinucleotide repeat that is expanded and unstable on Huntington's disease chromosomes. *Cell* 72:971-983.

Kalchman, M.A., Koide, H.B. , McCutcheon, K., *et al.* (1997). HIP1, a human homologue of S. cerevisiae Sla2p, interacts with membrane-associated huntingtin in the brain. *Nature Genetics* 16:44-53.

66

Kubo, T., Fukuda, K., Mikami, A., Maeda, A., Takahashi, H., Mishina, M., Haga, T., Haga, K., Ichiyama, A., Kangawa, K., *et al.* (1986). Cloning, sequencing and expression of complementary DNA encoding the muscarinic acetylcholine receptor. *Nature* 323, 411-6.

Lapham, E.V., Kozma, C., Weiss, J.O., (1996). Genetic discrimination: Perspectives of consumers. *Science* 274:621-4.

Maier, E., Meier-Ewert, S., Bancroft, D. and Lehrach, H. (1997) Automated array technologies for gene expression profiling *Drug Discovery Today* **2**: 315-324.

Maruyama, K. and Sugano, S. (1994) Oligo-capping: a simple method to replace the cap structure of eukaryotic mRNAs with oligoribonucleotides. *Gene Therapy.* 138: 171-174.

Munroe, D.J., Loebbert, R., Bric, E., Whitton, T., Prawitt, D., Vu, D., Buckler, A., Winterpacht, A., Zabel, B. and Housman, D.E. (1995). Systematic screening of an arrayed cDNA library by PCR. *Proceedings of the National Academy of Sciences of the United States of America* 92, 2209-13.

Pearson, W.R. and Lipman, D.J. (1988). Improved tools for biological sequence comparison. *Proceedings of the National Academy of Sciences of the United States of America* 85, 2444-8.

Polymeropoulos, M.H., Higgins, J.J., Golbe, L.I., et al (1996). Mapping of a gene for Parkinson's disease to chromosome 4q21-q23. *Science* 274:1197-1199.

Pulver, A.E., Wolyniec, P.S., Hausman, D., *et al* (1997). The Johns Hopkins University Collaborative Schizophrenia Study: an epidemiologic-genetic approach to test the heterogeneitiy hypothesis and identify schizophrenia susceptibility genes. *Cold Spring Harbor Symposium of Quantitative Biologyl* 61, in press.

Risch, N. and Botstein, D. (1996). A manic depressive history. *Nature Genetics* 12:351-353.

Rosen, D.R., Siddique, T, Patterson, D., Figlewicz, D.A., Sapp, P., Hentati, A., Donaldson, D., Goto, J., O'Regan, J.P., Deng, H.X., *et al.* (1993). Mutations in Cu/Zn superoxide dismutase gene are associated with familial amyotrophic lateral sclerosis. *Nature* 362 (6415): 59-62.

Schuler, G.D., Boguski, M.S., Stewart, E.A., Stein, L.D., Gyapay, G., Rice, K., White, R.E., Rodriguez-Tome, P, Aggarwal, A., Bajorek, E., *et al.*(1996) *A Gene Map of the Human Genome Science* 274: 540-546.

Sherrington, R., *et al.* (1995). Cloning of a gene bearing missense mutations in early-onset Alzheimer's disease. *Nature* 375:754-760.

Soares, M.B., Bonaldo, M.F., Jelene, P., Su, L., Lawton, L. and Efstratiadis, A (1994). Construction and characterization of a normalized cDNA library. *Proceedings of the National Academy of Sciences of the United States of America* 91, 9228-32.

Venter, J.C. (1993). Identification of new human receptor and transporter genes by high throughput cDNA (EST) sequencing. *Journal of Pharmacy and Pharmacology* 45 Suppl 1, 355-60.

Wilmut, I., Schnieke, A.E., McWhir, J., Kind, A.J. and Campbell, K.H. (1997) Viable offspring derived from fetal and adult mammalian cells. *Nature* **385**, 810-813.

USING THE MOUSE TO SIFT SEQUENCE FOR FUNCTION

EDWARD M. RUBIN

ABSTRACT

Libraries of all or part of the mammalian genome have generally been propagated in single cells and have been used as tools in gene discovery through *in vitro* analyses. We have expanded upon this concept by the creation of panels of Yeast Artificia ChromosomeYAC transgenic mice containing defined contiguous regions of the mouse or human genome. In these studies, we created a 2 megabase *in vivo* library of human chromosome 21q22.2 in a panel of YAC transgenic mice. Analysis of these animals revealed that a single YAC from this region was associated with specific behavioral and learning deficits. Through fragmentation of the YAC during the process of microinjection, we have been able to identify a human homolog to a *Drosophila* gene, *minibrain*, responsible for behavioral abnormalities in both flies and in these transgenic mice.

INTRODUCTION

The linking of sequence to biological function has become an increasingly significant bottleneck to harvesting the fruits of sequencing the human genome. Thus, the clear challenge facing the post-sequence world is the task of understanding how the genes discovered from DNA sequencing function in the living organism. Non-vertebrate model organisms, despite a proven utility in deciphering gene function at the organismal level, are inadequate for modeling many complex mammalian physiological traits (e.g., control of blood glucose, behavior, cardiac function, etc.). In contrast, the mouse, as a mammal, has been used to successfully model many complex human conditions. Ways in which the mouse has provided information on gene function include naturally occuring mouse mutations and the alteration of single genes through gene targeting and transgenesis. However, the ability to genetically engineer mice coupled with their utility for modeling many common human conditions is in part counterweighted by this organism's relatively long reproductive cycle, and small litters. This makes the mouse, in many ways, an inherently expensive and low-throughput reagent for deciphering the function of the ever increasing numbers of novel genes identified by the genome project.

Recent technical innovations in large insert transgenesis (Jakobovits *et al.*, 1993; Lamb *et al.*, 1993; Frazer *et al.*, 1995) and in creation of large deletions (Ramirez-Solis *et al.*, 1995; You *et al.*, 1997) have increased the number of genes which can be examined at once in a single animal. These technologies offer a new means for increasing the potential throughput of performing genotype/phenotype analysis in the mouse. The work described here concentrates on the development and analysis of panels of large insert transgenic mice propagating defined segments of the human genome. We call these panels of large insert transgenics, "*in vivo* libraries."

In vivo libraries: to identify genes contributing to phenotypes

Transgenic approaches have typically analyzed the function of discrete genes, or coding regions, one at a time (simplex). The *in vivo* library approach differs from these traditional strategies and involves making a series of large insert transgenic animals propagating DNA that covers a particular candidate region of the genome. The region may be chosen based on the presence of a mapped genetic locus that plays a role in a disease or physiological process. Because of the increased likelihood of including multiple genes as well as large genes, these large insert vectors (preferably YACs) maximize (multiplex) the amount of information that can be derived from a relatively limited panel of founder transgenic animals. Furthermore, because of the presence of normal *cis* regulatory elements, the use of genomic versus cDNA transgenes maximizes the likelihood of obtaining authentic patterns of expression and hence biological impact of the genes contained within the transgene.

An increased dose of 21q22.2 gives defined behavioral phenotypes

Our initial foray into the use of *in vivo* libraries concentrated on a 2 Mb region on the long arm of human chromosome 21 at 21q22.2. Studies of humans with partial trisomy 21 have suggested that an extra copy of this segment of chromosome 21 is sufficient to cause many of the phenotypic features of Down syndrome (Rahmani *et al.*, 1989). This has been supported by studies of mice trisomic for chromosome 16 syntenic with human chromosome 21 (Cox *et al.*, 1984). These animals display a number of phenotypes, including deficits in learning and memory. Hence, 21q22.2 was a particularly attractive region of the genome in which to search for genes affecting learning and memory.

To identify sequences contributing to the learning defects, we created an *in vivo* library based on the mapping studies, covering 2 Mb of 21q22.2 propagated as a panel of YAC transgenic mice (Smith *et al.*, 1995). Each member of the panel contained distinct 430-670 kb overlapping segments of the chromosome 21 region. The integrity of the YAC transgenes in the mouse genome was established by inter-alu PCR fingerprinting (Smith *et al.*, 1995). The copy number of the YAC transgenes was assessed and their integrity further confirmed by FISH (Smith *et al.*, 1997). This analysis showed that the copy number of the YAC transgenes ranged between 1 and 3 copies per mouse genome, and suggested that there was low level overexpression of the genes present on the YACs. Furthermore, RT-PCR analysis of at least one transcription unit on each one of the YACs suggested that the human genes were correctly transcribed in the foreign environment of the mouse genome (Cox *et al.*, 1984). For each YAC, two or more independent lines of animals containing the full-length transgene were created in order to be certain that any phenotypes observed in members of the library were replicated by the two independent lines. In this way, it could be verified that any phenotypes observed were most likely due to the extra copy of genetic information provided by the transgene, and not simply due to insertion effects of the transgene into an endogenous mouse gene.

Behavioral analysis of the *in vivo* library

Members of the library were subjected to detailed assays for learning and behavior in order to screen for genes which, when present at an extra dose, affected these phenotypes (Cox *et al.*, 1984). The Morris water maze, which has become a widely accepted test of learning and memory for small rodents, was employed. In comparison to non-transgenic controls, the most severe deficits on this test were shown by animals containing a 550 kb YAC, "152F7".

Fine mapping a gene involved in learning and memory

To localize the gene responsible for the learning and memory deficits of mice containing YAC 152F7, advantage was taken of the observation that fragmentation of the lengthy YAC DNA occurs in the handling during microinjection. This leads to a panel of animals that contains random fragments of the YAC, in addition to animals containing the full length unrearranged YAC. Animals containing random YAC fragments provide a valuable resource for ultra-fine structure mapping of genetic traits, since the number of break points obtainable as a result of the fragmentation are far more numerous than one could practically obtain using classical meiotic genetic mapping. Of the animals containing random fragments of YAC 152F7, animals containing a 180 kb telomeric fragment of the YAC showed learning and memory deficits that were indistinguishable from animals containing the full length YAC, whereas animals containing a complementary 390 kb centromeric fragment showed normal learning and memory. The fragmented YAC studies had thus reduced the interval containing the sequence(s) contributing to the learning defects from 570 kb to 180 kb.

DYRK is responsible for the learning and memory deficits

The only gene that appears to be present in the 180 kb telomeric region is DYRK (Cheng *et al.*, 1994; Guirema *et al.*, 1996; Kentrup *et al.*, 1996). Expression studies confirmed that this gene was overexpressed as a result of the transgenesis with both the full length YAC 152F7 and the telomeric fragment, and that the level of overexpression was consistent with the copy number of the transgenes.

The DYRK gene is over 100 kb and is the human homolog of the Drosophila gene *minibrain* (Tejedor *et al.*, 1995; Song *et al.*, 1996). This gene, *minibrain*, is of particular interest since mutations affecting the activity of the Drosophila homolog result in learning impairment in flies. The parallel consequences of altering expression of *minibrain* in flies and mice suggest that correct dosage of the *minibrain* gene is crucial for normal development of the nervous system.

The approach employed to identify a chromosome 21 gene affecting learning and behavior was in part based on the knowledge that an increased copy of genes from the 21q22.2 region played a role in many of the characteristic features of trisomy 21. An essential question remaining is whether this approach is generally applicable to other complex traits, not normally associated with increased gene dosage. This would require that increased dosage of responsible genes in the relevant candidate regions would affect the phenotype of interest. This clearly will not always be the case, but there are abundant examples suggesting that this sort of approach, especially when linked to sensitive phenotypic assays, will help localize genes contributing to complex traits. One such use of an *in vivo* library has been demonstrated by the successful fine mapping of quantitative trait loci involved in asthma (D. Symula and E.M. Rubin, unpublished results).

CONCLUSIONS

By using an *in vivo* library of large insert transgenic animals containing DNA from human chromosome 21q22.2, together with a functional assay, we have demonstrated that it is possible to sift through a large genomic region and identify distinct sequences affecting learning. Furthermore, the analysis of the *in vivo* library enabled the identification of a human gene (DYRK) from 21q22.2 causing learning defects in mice, the homolog of which (*minibrain*) had previously been shown to be involved in learning and memory in Drosophila. Together, these observations suggest that normal expression of DYRK or *minibrain* may be required for normal neural development.

71

The ability to link sequence to function in mammals has been far outstripped by production sequencing programs and the ability to map important candidate gene regions based on high throughput genotyping. The approach described here consists of screening in parallel the effect of many genes upon a single phenotype, in this case learning and memory, through the use of panels of large insert transgenic mice. In addition to the study described, the utility of *in vivo* libraries has also been demonstrated with regard to the biological annotation of genomic sequence data (Frazer *et al.*, 1997), and the fine mapping and eventual cloning of the mouse neurological mutation *vibrator* by *in vivo* complementation (Hamilton *et al.*, 1997). Just as new tools appeared in response to needs for scaling up gene mapping, it is likely that the bottleneck of linking genotype/sequence information to function will be eased with the development of other novel higher-"throughput" approaches employing the mouse.

REFERENCES

Cheng, J.-F., Boyartchuk, V. and Zhu, Y. (1994) Isolation and mapping of human chromosome 21 cDNA: progress in constructing a chromosome 21 expression map. *Genomics* **23**, 75-84.

Cox, D.R., Smith, S.A., Epstein, L.B. and Epstein, C.J. (1984) Mouse trisomy 16 as an animal model of human trisomy 21 (Down syndrome): production of viable trisomy 16 diploid mouse chimeras. *Developmental Biology* **101**, 416-424.

Frazer, K.A., Ueda, Y., Zhu, Y., Gifford, V.R., Garafolo, M., Mohandas, N., Martin, C.H., Palazzolo, M.J., Cheng, J.-F. and Rubin, E.M. (1997) Computational and biological analysis of 680 kilobases of DNA sequence from the human 5q31 cytokine gene cluster region. *Genome Research* **7**, 495-512.

Frazer, K.A., Narla, G., Zhang, J.-L. and Rubin, E.M. (1995) The apolipoprotein(a) gene is regulated by sex hormones and acute-phase inducers in YAC transgenic mice. *Nature Genetics* **9**, 424-431.

Guirema, J., Casas, C., Pucharcos, C., Solans, A., Domenech, A., Planas, A.M., Ashley, J., Lovett, M., Estivill, X. and Pritchard, M.A. (1996) A human homologue of *Drosophila minibrain (MNB)* is expressed in the neuronal regions affected in Down syndrome and maps to the critical region. *Human Molecular Genetics* **5**, 1305-1310.

Hamilton, B.A., Smith, D.J., Mueller, K.L., Kerrebrock, A.W., Bronson, R.T., van Berkel, V., Daly, M.J., Reeve, M.P., Nemhauser, J.L., Hawkins, T.L., Rubin, E.M. and Lander, E.S. (1997) The vibrator mutation causes neurodegeneration via reduced expression of PITPa: positional complementation cloning and extragenic suppression. *Neuron* **18**, 711-722.

Jakobovits, A., Moore, A.L., Green, L.L., Vergara, G.J., Maynard-Currie, C.E., Austin, H.A. and Klapholz, S. (1993) Germ-line transmission and expression of a human-derived yeast artificial chromosome. *Nature* **362**, 255-258.

Kentrup, H., Becker, W., Heukelbach, J., Wilmes, A., Schurmann, A., Huppertz, C., Kainulainen, H. and Joost, H.-G. (1996) Dyrk, a dual specificity protein kinase with unique strucutural features whose activity is dependent on tyrosine residues between subdomains VII and VIII. *Journal of Biological Chemistry.* **271,** 3488-3495.

Lamb, B.T., Sisodia, S.S., Lawler, A.M., Slunt, H.H., Kitt, C.A., Kearns, W.G., Pearson, P.L., Price, D.L. and Gearhart, J.D. (1993) Introduction and expression of the 400 kilobase *precursor amyloid protein* gene in transgenic mice. *Nature Genetics* **5,** 22-30.

Rahmani, Z., Blouin, J.-L., Creau-Goldberg, N., Watkins, P.C., Mattei, J.-F., Poissonnier, M., Prieur, M., Chettouh, Z., Nicole, A., Aurias, A., Sinet, P.-M. and Delabar, J.-M. (1989) Critical role of the D21S55 region on chromosome 21 in the pathogenesis of Down syndrome. *Proceedings of the National Academy of Sciences, USA* **86,** 5958-5962.

Ramirez-Solis, R., Liu, P. and Bradley, A. (1995) Chromosome engineering in mice. *Nature* **378,** 720-724.

Smith, D.J., Stevens, M.E., Sudanagunta, S.P., Bronson, R.T., Makhinson, M., Watabe, A.M., O'Dell, T.J., Fung, J., Weier, H.-U.G., Cheng, J.-F. and Rubin, E.M. (1997) Functional screening of 2 Mb of human chromosome 21q22.2 in transgenic mice implicates *minibrain* in learning defects associated with Down syndrome. *Nature Genetics* **16,** 28-36.

Smith, D.J., Zhu, Y., Zhang, J.-L., Cheng, J.-F. and Rubin, E.M. (1995) Construction of a panel of transgenic mice containing a contiguous 2-Mb set of YAC/P1 clones from human chromosome 21q22.2. *Genomics* **27,** 425-434.

Song, W.-J., Sternberg, L.R., Kasten-Sportes, C., Van Keuren, M.L., Chung, S.-H., Slack, A.C., Miller, D.E., Glover, T.W., Chiang, P.-W., Lou, L. and Kurnit, D.M. (1996) Isolation of human and murine homologues of the *Drosophila* minibrain gene: human homologue maps to 21q22.2 in the Down syndrome "critical region". *Genomics* **38,** 331-339.

Tejedor, F., Zhu, X.R., Kaltenbach, E., Ackermann, A., Baumann, A., Canal, I., Heisenberg, M., Fischbach, K.F. and Pongs, O. (1995) *minibrain*: a new protein kinase family involved in postembryonic neurogenesis in Drosphila. *Neuron* **14,** 287-301.

You, Y., Bergstrom, R., Klemm, M., Lederman, B., Nelson, H., Ticknor, C., Jaenisch, R., and Schimenti, J. (1997) Chromosomal deletion complexes in mice by radiation of embryonic stem cells. *Nature Genetics* **15,** 285-288.

EXPLOITATION OF SEQUENCE INFORMATION

FROM SEQUENCE TO STRUCTURE

FRANK E. BLANEY

SmithKline-Beecham Pharmaceuticals, Computational and Structural Sciences, New Frontiers Science Park (North), Third Avenue, Harlow, Essex, CM19 5AW, UK

ABSTRACT

The explosion of sequence information from the various genomic efforts has far outweighed the number of actual protein structures determined by experimental means. Thus it is increasingly important that methodology be developed which can accurately predict the structure of proteins from their sequence. This is often called the "protein folding problem" and this chapter will briefly review the various computational methods available for this. At its highest level, structure prediction is based on the technique of homology modelling but *de novo* methods such as threading, reduced energy functions and simplistic lattice models have all been used and have proven successful in various cases. It can be concluded that while the protein folding problem is still to be solved, significant advances have been made and the ultimate goal of accurate structure prediction still remains an achievable target

INTRODUCTION

The veritable explosion of sequence data arising from the various genomic projects has given researchers a golden opportunity to further understand the nature of the proteins which control our living state. In addition to a further insight into genetic disorders, there is the possibility of gaining a greater knowledge of poorly understood disease states and exploiting this knowledge in the design of novel therapeutic treatments.

One perceived problem is that gene sequence data alone is not enough to help, for example, in the design of novel drug molecules. It is relatively straightforward to search for similarities to known proteins at the DNA or amino acid level. If such similarities are found then this may go a long way towards the understanding of the function or even structure of the unknown protein. However, such a scenario is seldom the case and this has tempered the expectations of many scientists working on proteins as to the real value of such a wealth of data. Indeed, there is seen to be something of a cultural divide between molecular biologists and protein structural scientists, a fact which was highlighted in a recent editorial article in Nature Structural Biology (Editorial, 1997). Comments such as Stephen Bryant's "What do molecular biologists fear most? 3 letters PDB !!" serve well to illustrate the problem.

EXPERIMENTAL DETERMINATION OF PROTEIN STRUCTURE

In essence, the main problem may be outlined as "Given a protein sequence, how can this information be used to gain knowledge of its three-dimensional (3-D) structure and perhaps even more importantly, its biological function?" Only with this information can full use be made of modern techniques such as structure based drug design. One obvious way forward is to express the protein and then determine its 3-D structure experimentally. There are of course many problems associated with the expression of proteins which are beyond the scope of this article. However, even when a protein has been successfully expressed in sufficient quantities, there are still many problems which can arise in structural determination. The traditional technique of X-ray crystallography only applies to proteins which can be crystallised and hence excludes, for example, the majority of membrane bound receptors. Even soluble proteins themselves are often difficult to crystallise. The extent of the problem is illustrated in Figure 1 which shows the number of structures solved [as determined by the number deposited in the Brookhaven database (Bernstein *et al.* 1977)] vs. the number of sequences identified over the period from 1985 to 1996. As of April, 1997, some 5400 structures were deposited in the PDB (Brookhaven) database. There are many more "confidential" structures solved e.g. by pharmaceutical companies but even with these, the number of available sequences is increasing at a much greater rate than their structural elucidation.

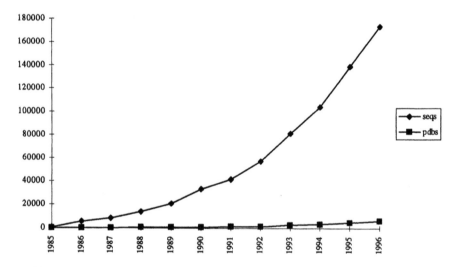

Figure 1. The number of sequences vs. number of structures from 1985-1996.

In addition to X-ray crystallography, other techniques have been used in recent years to solve protein structures. One of the most important of these is multidimensional nuclear magnetic resonance (NMR) spectroscopy (Evans, 1995). This has the obvious advantage that, being in solution, there are no problems with crystallization. Multidimensional nmr relies heavily on the use of the nuclear Overhauser effect (NOE) (Neuhaus and Williamson, 1989). Protons which are distant in terms of sequence are often close together in actual space as a result of the protein folding. Irradiation of such a proton signal can cause an enhancement of those which are spatially close to it, and the intensity of the observed enhancement is related to the actual distance between them. Thus by collecting all the observable distances between identified atoms one can use techniques such as *distance geometry* (Crippen and Havel, 1988) or *constrained molecular dynamics* (Zuiderweg *et al.* 1985) to iteratively determine the 3-D structure. With these techniques, of course, there is not always a unique solution and one often therefore gets an ensemble of structures. NMR has been applied with considerable success to a large number of small proteins. Further details may be found for example in one of the excellent textbooks on the subject (Evans, 1995).

Another technique which has been successfully applied to the solution of protein structures not amenable to X-ray crystallography, is cryoelectron diffraction. This in theory, has the same potential resolution as X-ray diffraction. The main problem is that the very high energy sources involved tend to destroy the molecules; hence the need for low temperatures and short sampling times. The method has been particularly useful in the determination of membrane bound receptors such as bacteriorhodopsin (Henderson *et al.* 1990) and bovine (Schertler *et al.* 1993) and frog (Unger *et al.* 1997) rhodopsins. If expression levels can be increased, then it is expected that this technology will play an important role in the determination of other 2-D structures such as membrane bound transporter and ion channel proteins.

Other spectroscopic techniques such as circular dichroism (CD) and mass spectrometry have been applied to protein structural elucidation but, despite advances in the technology/methodology, it is unlikely that the experimental methods will ever become routine enough to keep pace with the sequence information. For this reason many researchers have developed and employed theoretical computational techniques to answer the question, "Can an unknown protein structure be predicted from its sequence?" This problem, often known as the ***protein folding problem***, has been the subject of extensive research over the last two decades.

OBSERVATIONS ON PROTEIN STRUCTURE

Despite the almost infinite number of possible conformations, proteins have for a long time been recognised to fold into well-defined "secondary structural elements". They are classified primarily by the values of their backbone dihedral angles (Brandon and Tooze, 1991) (the so-called ϕ or $C\alpha$-N_{amide} angle and ψ or $C\alpha$-C_{amide} angle). There are four broad groups of these:

- Helix   (average $\phi = -55°\pm20°$ average $\psi = -55°\pm20°$
- Extended (Sheet)   (average $\phi = -120°\pm45°$; average $\psi = +130°\pm30°$
- Turn   (various values depending on turn type)
- Loop (Coil)

The first three of these are well defined geometrically. Turns refer to a reversal of direction around four amino acids and are further subdivided into a number of different types according to the ϕ and ψ angles of the central two residues. The loop or coil category has no defining 3-D structure and simply refers to all the lengths of sequence that cannot be classified into one of the other categories. It has been found however that based on analysis of the PDB database, most amino acids have well defined "allowed" values of ϕ and ψ angles. This fact was first noted by Ramachandran who published the plots of allowed ϕ vs. ψ angles which now bear his name (Ramachandran and Sassiekharan, 1968).

These various secondary structural elements then pack together to give distinct domains or "folds", usually classified as all α, all β or mixed α/β (Brandon and Tooze, 1991). In larger proteins, the structure may consist of several domains packed together. It has been recognised for a long time that if one could accurately predict these secondary structure elements, then the problem of folding these into the native state of the protein is greatly simplified.

SECONDARY STRUCTURE PREDICTION

Many methods have evolved over the last twenty years for the computational prediction of secondary structure from primary sequence. Among the best known are those of Chou and Fasman (1974), Lim (1974) and GOR (Garnier *et al.* 1978). More recently combination methods such as the WASP technique (Boscott *et al.* 1993) have been used but despite all these "improvements", secondary structure prediction seems to have reached a constant level of at best around 70% accuracy.

The earliest method to be described was that of Chou and Fasman back in 1974. In this, a statistical analysis was carried out on 15 members of the PDB database and propensities (P_x) were assigned to each amino acid, as to its likelihood to exist in an α-helix (P_a) and β-sheet (P_b). If the value of the propensity was below 1, then the residue was likely to be a strong breaker of that type of secondary structure. Thus glutamic acid with a P_b=0.26 stongly disfavors β-sheets whereas its P_a value of 1.53 means that it is frequently found in α-helices. Empirical rules were developed to predict over a given length of sequence, when secondary structural elements would start and finish. The method is widely used, partly because of its ready availability, but although the original authors claimed an accuracy of 77%, a more recent study by Kabsch and Sander (1983) suggested that the method does no better than 50%.

In GOR, Garnier and co-workers attempted to overcome some of the shortcomings of the Chou-Fasman method by including the effects of neighboring residues. They used the more common four state model (ie. helix, sheet, turn and coil) with a window length of 17 residues. The influence of each residue at positions up to 8 on either side of the central residue was summed up, to give a final prediction of the state. As reported in the review by Kabsch and Sander (1983), using a set of 62 proteins and a three-state prediction (helix, sheet and other), GOR scored slightly better than Chou-Fasman at around 56%.

More recently, neural networks have been used as an alternative approach to secondary structure prediction (Rost and Sander, 1993b). The average success rate of these however is not much better at ~65%. The main problem is that all the methods above use information local to the sequence but secondary structure is often controlled by non-local effects as a result of the folding of the protein chain. Without taking these long range effects into account, the likelihood of achieving predictions greater than 70% is very small. Further proof of this came from Kabsch & Sander (1984) who showed that in an analysis of pentapeptide sequences within the PDB, the same sequence could often adopt totally different secondary structures (see Figure 2). This should also act as a warning to those looking for homology at the sequence level!

Despite the failure rate of the older methods described above, it has been found that the inclusion of multiple sequence alignment data within a neural net method such as the

PHD program (Rost and Sander, 1993a), has led to an average improvement of 6%. It is commonly believed that PHD or other multiple sequence based programs are now the best secondary structure prediction methods available.

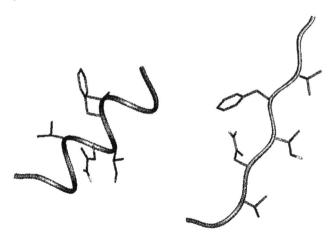

Figure 2. Secondary structure of the sequence, Val-Asn-Thr-Phe-Val. In erythrocruorin (left) it is in an α helix whereas in ribonuclease (right), it is a β-strand.

The folding of secondary structural elements into the native state has been discussed by numerous authors. Many have used energy based methods such as molecular dynamics or Monte Carlo while others have relied on rule based techniques. Some of these will be discussed in more detail later. The overall principle is nicely illustrated by Cohen (Curtis et al. 1991) in the construction of their model of IL4. Here secondary structural units such as the helices were predicted by the methods of Chou-Fasman and GOR, and these were then packed so as to maximize the favorable energy.

PREDICTIONS BASED ON OBSERVATIONS OF PROTEIN FOLDING

Examination of available structures in the PDB database has led many people to observe that proteins with similar sequence identity and/or function will tend to adopt the same fold. What is perhaps more suprising is the observation that proteins with little or no sequence homology and with different functions, can also adopt identical folds. These are often called fold superfamilies, examples being the TIM barrels and the four helix bundles.

Indeed a much smaller number of overall protein folds has been observed than would be expected from the number of structures solved. These observations have been the basis for two major techniques used in theoretically derived protein structure predictions, *viz* *homology modelling* and *inverse folding* or *threading*.

Homology Modelling

Given a gene or a translated amino acid sequence, the first step in any analysis of the protein is a multiple sequence alignment, i.e. a comparison of that sequence with other known sequences in either proprietary or public domain databases. Many programs such as BLAST (Altschul *et al.* 1997), CLUSTALW (Thompson *et al.* 1994), etc, are available for this procedure. Alignments can be described in terms of sequence identity and if an unknown sequence is above a certain threshold in sequence identity (often quoted as ca. 30%), then it is reasonable to assume that the proteins are related. If the 3-D structure of one or more of these "known" sequences is available, then models can be built of the related unknown sequence using the technique of homology modelling. It is generally accepted that this is the most accurate way to model an unknown protein from sequence information alone but of course, homology modelling is unsuitable if no known 3-D structures of related proteins are known. Early examples where this has been particularly successful are the homology models of the aspartyl proteases, renin (Hemmings *et al.* 1985) and HIV protease (Pearl and Taylor, 1987), both of which were modelled before the crystal structures were determined. The technique is now routine and has in fact been totally automated in several programs. It consists of a number of well defined steps:

- A multiple sequence alignment of the unknown sequence is carried out with related proteins of known structure.

- Overlap the known 3-D structures (if there is more than one) based on the sequence alignment. Multiple 3-D structures give more confidence in the identification of the structurally conserved regions (SCRs) below.

- Identify the core regions which are structurally conserved. These "SCRs" are usually in helical or extended sheet regions of the proteins and are connected by the variable loop regions.

- Construct the core backbone of the unknown protein based on the sequence alignment with the SCRs.

- Build the unknown (often variable loop) regions. For shorter loops this is most often done by searching libraries of known loops. Alternatively, energy based methods such as constrained molecular dynamics and minimisation can be used.

- Add the amino acid sidechains with sensible values

- Refine the structure (usually by minimization or molecular dynamics).

- Check on the "goodness" of the model, ie. check for the presence of abnormal protein features.

Aligning the sequences. It is generally agreed (Lozano *et al.* 1997) that the most important stage in any homology modelling exercise is the initial sequence alignment; an error at this point can often have a dramatic influence on the final structure. Unfortunately, several problems can be associated with automated sequence alignment. Any of the commonly used programs already mentioned will align two proteins and give a best result, even if in reality this is totally coincidental. The score which is given as a measure of this alignment can be based on percentage identity or calculated with one of the many mutation matrices available, and then compared to the score obtained from random sequences. Unfortunately, the lower the similarity, the greater is the statistical probability of a match being simply due to chance. For a protein of 80 residues or more, a similarity of less than 25% identity may not be significant or better than a random match (Feng and Doolittle, 1987).

Multiple sequence information may help in improving the alignment by identifying structurally important features which are conserved. These would not show up necessarily when only two sequences are aligned. Conserved cysteine residues may be found for example, when they are involved in disulphide bonds. The mutation to a proline residue in a helix may cause disruption as may the mutation to a polar residue in a conserved hydrophobic core position of a globular protein. Active site residues are often highly conserved across a family but alignment of two sequences alone would not necessarily show this up. Automatic alignment methods thus do not necessarily do the best job at predicting similarity. Use should always be made of multiple sequence information where available and manual realignment should be carried out if the automated result is felt to be wrong. After all the human eye is a very good implement for pattern recognition, particularly where use can be made of colour-coding (Morris, 1988) in the viewing of alignments.

Structurally conserved regions. Observations on the 3-D structure of the known homologues can be used with the multiple sequence alignments to identify the structurally conserved regions (SCRs). These are typically, areas of well defined secondary structural element types, ie, α-helices or β-sheets, which are connected by conformationally unknown stretches of sequence. Wherever possible, multiple known structures should be superimposed to define the conserved 3-D backbone. This can be a useful addition to the use of multiple sequence alignments alone as it is not biased by the restrictions of the scoring matrices. The commonly used modelling programs, such as Quanta (QUANTA), InsightII (INSIGHT), Sybyl (SYBYL), etc all have algorithms for carrying out this step routinely. It is also possible in these programs, with a 3-D alignment, to generate an average structure or to copy different SCRs from different structures. This may have some advantages over any single member, eg. where there is greater local sequence identity in specific SCRs with the unknown sequence.

Loop searching. The most difficult aspect of homology modelling and the one where most advances are likely to be made in the future, is the construction of the unknown loop regions which span the SCR framework defined above. Ignoring four residue β-turns (Wilmot and Thornton, 1988), for which the average ϕ–ψ values of different amino acids at each position have been well established, short loop geometries are probably best evaluated using energy-based conformational searching. This will generate an ensemble of configurations (Moult and James, 1986) which can ranked by energy. Using standard force field programs this rapidly becomes impossible with loops of seven residues or more. However the ICM program (Abagyan et al. 1994) or the conformational poling algorithm (Smellie et al. 1995) can extend this limit, and the former method has been used successfully to look at loops of up to 25 amino acids. Distance geometry (Crippen and Havel, 1988) represents an alternative technique for the generation of conformations. Another method which is extensively used for longer loops is to search a database of known loop geometries (generated from the PDB database). Standard databases containing the necessary information are again available with the commonly used modelling programs mentioned above. Sequences of the same length as the required loop and the correct geometry to connect to the two neighboring SCRs are extracted. The best loop can be chosen on the grounds of sequence similarity and/or geometry and energy minimised at a later date.

A disadvantage of the common loop searching methods is that they do not take account of the environment around that loop, i.e. non-local effects. An important database advance in 3-D protein searching has been reported by Sander, Vriend and Stouten (Vriend et al. 1994) at EMBL, Heidelberg. Their program, SCAN3D, maintains the usual relational database paradigm where possible, but holds in addition, property profile information over whole sequences and a three level hierarchical structure of information on the protein / residue / atom basis. It is not possible in this report to go into greater detail but the program certainly seems to hold much promise. It is possible, for example, to retrieve loop conformations from proteins where the neighbouring residue environments are also specified.

Sidechain modelling. Once the final backbone structure has been generated, the various sidechains must then be added in sensible geometries. An examination of the PDB database has shown that the various sidechain χ-angles tend to exist in well defined discreet ranges. This is because rotations can often cause the sidechain to clash with the backbone or to adopt a local eclipsed conformation. For this reason so called *rotamer libraries* have been defined by several groups (Ponder and Richards, 1987; Summers *et al*. 1987). These define the preferred sidechain dihedrals in each of the standard α-helix, β-sheet or coil backbone regions. They are often a useful starting point for, or alternative to, energy minimizations. Their application involves a systematic search for all possible rotamer values. This can lead to combinatorial explosion problems with larger proteins but a sensible choice of the order in which the library is applied can greatly reduce this; i.e. look at residues which are both spatially and sequentially adjacent, at the same time, rather than going through the whole sequence, residue by residue.

Refinement and protein checking. Once a homology model has been built, the final refinement is performed using molecular mechanics minimization or molecular dynamics. Checks on just how good a protein structure is, rely on aspects such as goodness of backbone dihedrals (not too many positive angles in the Ramachandran map), estimates of buried hydrophilic or exposed hydrophobic residues which are not generally found in globular proteins and searches for cavities. Programs such as Procheck (Laskowski *et al*. 1993), Qpack (Gregoret and Cohen, 1990) and the Quanta Protein Health (QUANTA) option and are all readily available to perform this task.

The whole procedure of homology modelling has been well documented and examples too numerous to list here, have been described in the scientific literature. The technique has become so routine in fact that it has been automated in a number of programs which are now commercially available. The best known of these are Composer (Sali *et al*. 1990; Blundell *et al*. 1989), Modeller (Sali and Blundell, 1993) and Consensus (Havel, 1993). There is also a service available on the world-wide web, known as SwissModel (Peitsch, 1996), whereby upon submission of a sequence, a homology model will be returned. While these are general homology modelling packages, similar programs have also been developed for particular classes of proteins such as monoclonal antibodies (Martin *et al*. 1989). Despite this automation however, homology modelling is still vitally dependent on the initial sequence alignment. It is imperative that this should be checked manually.

Structural similarity from sequences with no identity

As was mentioned earlier, the suprising observation has been made that proteins with little or no sequence homology and different functions can adopt identical folds. A classic example of this is the family of α/β proteins known as the TIM barrels, which consist of eight β-strands surrounded by eight α-helices. This motif is found in a wide variety of proteins (eg. triosephosphate isomerase, glycolate oxidase, etc. Figure 3) with different functions and no sequence identity. These structural similarities probably reflect an evolution from a much smaller number of primordial folds. It is widely believed in fact that despite the diversity of protein sequences now known, there is a finite limited number of folds (Bowie *et al.* 1991). If the number of folds is indeed limited, then it is conceivable that X-ray and NMR will eventually identify <u>all possible</u> folds.

Sequences with NO Identity in Sequence or Function

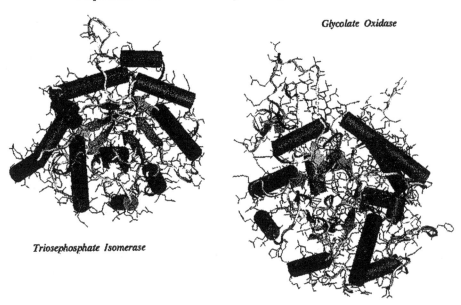

Figure 3. Two typical TIM Barrel proteins.

For this reason searchable databases such as SCOP (Murzin *et al.* 1995) and CATH (Orengo *et al.* 1997) have been developed which contain all existing known folds. It follows that given a sequence of unknown structure and knowing all folds, it should be possible to identify which fold best fits the sequence. This is known as *inverse folding* or *threading*.

The problem of how to check if a given sequence is compatible with a known fold was actually first posed by Ponder and Richards (1987) but credit for the real pioneering research into protein threading must be given to two groups, viz. Finkelstein & Reva (1991) and Eisenberg's group (Bowie *et al.* 1991) at UCLA.

Finkelstein & Reva started with an idealised β-barrel structure and matched various protein sequences against this. In their process no gaps were allowed, which was a limitation. However they were able to correctly predict the structures in three out of four cases. The method only took into account the interactions of the core residues, which is presumably acceptable for β barrel structures. It cannot be readily extended to other types of proteins.

In the method developed by Eisenberg and co-workers, each amino acid position in a protein is described in terms of a structural profile. Thus for each amino acid, the following properties are determined: the total surface area of side chain buried, the fraction of side chain area covered by a polar atom or water, and the local secondary structure (ie. α, β or other). Eighteen classes of sidechain are thereby defined. For each secondary structural type these are "E" for exposed, "P" for partially buried, and "B" for buried with the P-class split into P1 for less polar and P2 for more polar and the "B" class split into B1, B2 and B3, in order of increasing polarity. A library can be built up of proteins in a given fold, which describes as 1-D strings, the likelihood of any amino acid existing in the corresponding 3-D position of that fold. This profile can be used in the same way as a Dayhoff-type mutation matrix (Dayhoff *et al.* 1972) to carry out alignment of the test sequence with a database of known folds, and introducing gaps in a similar manner. The unknown sequence is scored against the various folds, with a high score indicating that the sequence is likely to exist in that fold. This has been tested for many cases, eg. in a search against the Actin fold, a hit was obtained for the 70 kD bovine heat shock protein, which has virtually no sequence identity to it. The two proteins do however have strikingly similar 3-D structures (Figure 4).

Results from Eisenberg's Threading Method

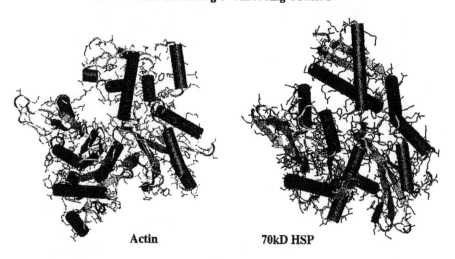

Actin   70kD HSP

Figure 4. The two proteins above were identified by threading to have similar folds.

The main problem with the method is that mutations in the sequence are only scored locally and thus do not take account of the 3-D environment (because it is defined by a 1-D sequence). Thus the reversal of an ion-pair, which would in reality be a favorable double mutation, would be predicted as being unfavorable.

An alternative threading strategy developed by Jones and co-workers gets over this problem (Jones *et al.* 1992). Here the fold was only defined as a backbone chain in 3-D space, with all sequence information being ignored. A library of all folds was thus created. The test sequence is then optimally fitted to each library fold, allowing indels in the loop regions only. The energy of each threading is evaluated as the sum of pairwise interactions, using potentials derived from statistical analysis of the PDB database. These potentials are simple atom pair distance functions. A cutoff of 10Å is applied for long range interactions and a residue solvation term is added to compensate. When tested, the method was able to identify C-phycocyanin as a globin fold despite the fact that there are only 14 amino acid identities out of 174 between it and sperm whale myoglobin (Figure 5).

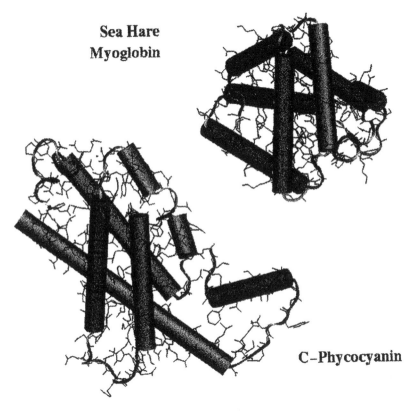

Figure 5. Using the threading algorithm of Jones and co-workers, these two proteins were predicted to have similar folds, despite a very low sequence identity.

A number of other approaches have been published (Sippl and Weitckus, 1992; Godzik *et al.* 1993; Maiorov and Crippen, 1992; Bryant and Lawrence, 1993; Ouzounis *et al.* 1993). All of these use highly simplified energy functions to score against known target structures.

There is no doubt that the threading approach to protein structure prediction has been valuable in many cases. Where no sequence identity can be established with known structures, it is possible to use threading instead of standard alignment methods as a starting point to homology modelling. The problem arises if the unknown structure corresponds to a fold which has yet to be discovered. In this case the use a totally *de novo* approach must be considered when folding the sequence.

ALTERNATIVE STRATEGIES TO PROTEIN FOLDING

The whole problem of protein folding has been the subject of extensive research for over two decades and it must be admitted that it still remains unsolved. Many of the methods used are based on Anfinsen's thermodynamic hypothesis (Anfinsen, 1973) that the native folded protein exists in the global minimum free energy state. Methods based on energy calculations however suffer from Levinthal's paradox (Levinthal, 1968) of how a long polypeptide chain searches the enormous number of conformational degrees of freedom to get to the native state. He has estimated that to search the conformational space of a small 100 amino acid protein using the forcefield methods described below would take 10^{50} years! While some groups have continued to use "full energy" methods, the majority of research has centered around the application of reduced (simplified) energy functions. These, together with the applications of prior secondary structure predictions, "U-turn" prediction algorithms (Kolinski *et al.* 1997) and simple idealised models of globular proteins such as regular lattice grids, have led to a host of structure predictions, many of which have been remarkably successful. A full description of this work is beyond the scope of this chapter but some details will be discussed below.

Full energy approaches

In principle the total energy of a system should be calculable by *ab initio* quantum mechanical methods but the computational resources to carry out such calculations on small peptides, let alone proteins, is not going to be available for a long time. In the field of macromolecules therefore, energy calculations have been based on the so called *molecular mechanics* method. This treats a molecule essentially as a system of "balls and springs" with the total energy of the system being calculated as the sum of various harmonic functions for bonded atoms, and other empirical functions for nonbonded atom pairs. A typical form of this is shown below. The k_b, k_θ and k_ϕ terms are force constants associated with bond stretching, angle bending and dihedral rotation respectively, A_{ij} and C_{ij} are the van der Waal repulsive and attractive terms between nonbonded atom pairs at a distance r_{ij} apart and q_i is the charge on atom i. The r_0 and θ_0 are equilibrium values of the appropriate bond length and bond angle respectively. The various force constants and equilibrium values etc, are held in associated databases known as *force fields*.

$$E_{total} = \sum k_b(r - r_0)^2 + \sum k_\theta(\theta - \theta_0)^2 + \sum |k_\phi| - k_\phi \cos(n\phi) + \sum_i \sum_j \frac{A_{ij}}{r_{ijr}^{12}} - \frac{C_{ij}}{r_{ij}^6} + \frac{q_i q_j}{r_{ij}}$$

The calculation of the energy of a molecule using these methods is many thousands of times faster than with quantum mechanics, yet to systematically explore conformational space for a protein would still be impossible.

One notable exponent of the full force field method is Scheraga at Cornell. Rather than systematically search around all the possible local minima on such a vast energy surface, he and his group have for the last 20 years, concentrated on mathematical solutions to the multiple minima problem, to get to the global minimum (native folded structure). Using the ECEPP force field (Browman *et al.* 1975), they have explored many methods and have settled more recently on the diffusion equation algorithm as giving the best results. This has performed well for oligopeptide structures (Kostrowicki and Scheraga, 1992). In a more recent alternative approach, they have used a united residue approximation (i.e. having very simple inter-residue rather than inter atom potentials) to initially search conformational space extensively. They then convert these results to real backbone conformations and optimise the H-bond network, before reverting to the full atom ECEPP force field, with an added hydration term. Using this combined approach they were able to correctly fold the 36 residue avian pancreatic polypeptide (Liwo *et al.* 1993).

An alternative approach has been developed by Abagyan. Rather than optimise geometry in cartesian space, ie. by incremental movement of individual atoms, he has used a biased Monte Carlo approach, in dihedral space only, in his ICM program (Abagyan *et al.* 1994). This force field also contains pseudoentropic terms to deal with sidechain flexibility and a solvation approximation (normally ignored in other force fields). The ICM method has been successful in cases where it has been applied but it is limited to probably 25 amino acids maximum. It is therefore very good for loop refinement rather than whole structures.

Reduced energy functions

As has been demonstrated above, the full energy functions have essentially only been useful for small polypeptides and are still not extendable to real proteins of interest. To overcome the combinatorial explosion associated with conformational searching using such functions, much effort has gone into simplification both of the functions and their associated force fields.

One method which is obvious is to assume that all bond lengths and angles are ideal. The backbone ω-dihedrals (i.e. amide bonds) are also kept fixed at $180°$. The bond stretching and angle bending terms of the equation can then be ignored. While the problem is now reduced to a search in dihedral space only, this does not give as much reduction in the computation as might be expected. The main effort in the calculation is in the non-bonded van der Waal (VDW) and electrostatic terms which rise with the square of the number of atoms. As was seen with the ICM method, realistic searches of total conformational space are restricted to 25-30 amino acids. To simplify further it is necessary to reduce the total number of atoms involved in the problem.

Much use has been made of the "united atom" approach. Here in its most common form, it is assumed that only polar hydrogens are explicit. All methyl, methylene and methine groups are then considered as single atoms with appropriately expanded VDW radii, force constants, etc. Using this approximation, the effective number of atoms can be cut by more than 50%. An early example of this was the UNICEPP program of Scheraga (Browman *et al.* 1977), which was cut down from his ECEPP force field. Most other commonly used force fields such as Charmm (Brooks *et al.* 1983) and Amber (Weiner *et al.* 1984) have associated united atom terms.

Another approach is to replace the all atom description by appropriate residue-based terms. In extreme cases this has meant using only a few points for each residue! Crippen, for example, has devised a contact-only potential form involving backbone N and O and sidechain Cβ atoms (Crippen, 1991). Such approximations are too gross however, unless they are used as a starting point to search conformational space extensively before replacing the per residue descriptors with real atoms (Liwo *et al.* 1993).

The method described by Osguthorpe (1997) serves to illustrate an intermediate approach. The backbone of each residue is represented by one sphere or pseudoatom whereas sidechains are represented by up to three. Alanine, valine, isoleucine, serine, threonine and proline have one sphere; leucine, histidine, aspartate, glutamate, asparagine, glutamine, cysteine and methionine have two; and phenylalanine, tyrosine, tryptophan, lysine and arginine have three. This reflects the shape of the sidechains better than would be possible with a single pseudoatom. Unlike others who have used reduced sidechain representations, Osguthorpe has developed a full potential force field including novel virtual stretch, bend and dihedral terms for the pseudoatoms, as well as the more conventional non-bonded term. Interestingly he has chosen to include a secondary structural element stabilisation term and a solvation term, which includes hydrophobic and dielectric charge effects. Conformational sampling is via a regular molecular dynamics algorithm.

Several groups have taken the approach of keeping an all atom backbone but searching in a limited discrete ϕ–ψ dihedral space. Gunn and co-workers examined the folding of a number of α proteins using a total of 18 allowed ϕ–ψ pairs commonly found in the PDB database (Gunn *et al.* 1994). The sidechains are represented by Cβ atoms only and the potential is of an interresidue type and distance function, based on a hydrophobic residue scale. Some prior knowledge of secondary structure was assumed here. The conformational sampling was done by a complex combination of loop lookup tables, genetic algorithms, simulated annealing and Monte Carlo. In a test case of 146 residues of myoglobin, the final structure was within 6.2Å of the native PDB state.

Yue and Dill (1996) used a united atom approach again with a discrete (up to 6 per amino acid) number of ϕ–ψ pairs. However by using a very simplified potential function involving only a maximisation of the number of hydrophobic contacts, they were able to completely search conformational space using their **constraint-based exhaustive searching** algorithm (Yue and Dill, 1995). This was originally developed during their work on lattice models (see below). On testing it found native-like conformations for crambin, avian pancreatic polypeptide, melittin and apamin, with no prior knowledge of structure.

A final approach which should be mentioned is due largely to the work of Sippl. Using an analysis of the PDB, he has developed a set of potentials of mean force which are distance functions for different atom type pairs (Sippl, 1990). The problem of searching conformational space is reduced by establishing a set of admissible conformations of segment length L (L is typically 5-7), taken from the database. The unknown sequence is mounted over each of these segments, an energy computed and the energies then sorted. Ensembles of low energy segments are then clustered to give one of four possible states from which starting points can be further evaluated.

Lattice models

In the whole field of protein folding simulation, two overriding theories have emerged. (1) local interactions determine the protein fold, i.e. local sequence determines the preformation of secondary structural elements such as α-helices, β-strands and turns, which then go to form the final tertiary structure. This is essentially the basis of many of the approaches which have been discussed already. In the second theory (2) it is non-local specific interactions which cause a collapse of the protein chain with subsequent formation of the secondary structural components. In particular, the dominance of hydrophobic residues in the core of globular proteins, has led to the conclusion that a hydrophobic collapse is the driving force in the folding of these structures. This is born out somewhat in consideration of the role of chaperone proteins. These can promote protein folding by providing a suitable environment in which amino acid chains can fold and unfold until they quickly gain a maximum "hydrophobic collapse", akin to a native globular protein fold. The use of simple exact lattice models was developed to simulate this latter scenario and has met with considerable success in the field of *ab initio* small globular protein folding.

In its simplest form, a cubic lattice (Figure 6) can be used to represent a protein with each amino acid situated at a grid point. Grid points can either be unoccupied or have a single occupancy. Various algorithms have been generated to fold an amino acid chain onto the grid and simple energy functions are used to assess the goodness of the fold. Forcing the single occupancy rule avoids chain crossing and also implies an inbuilt inter-residue VDW repulsion. The used of a fixed lattice also implies that there are no angle or dihedral terms in the energy function. Because of all the simple assumptions arising from the uniform geometry, the simplified energy function and the restriction on folding pathways, the sampling/search speed should be much higher than with standard geometry-energy based methods.

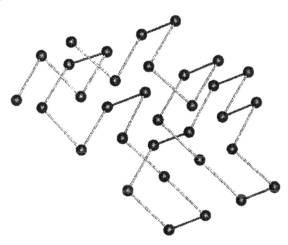

Figure 6. A cubic grid representationof two adjacent short α-helices.

Two typical cubic lattice models have been described by Covell (Covell, 1992; Covell, 1994). Analysis of known protein structures has shown that the average Cα-Cα distance is fairly invariant at 3.8 Å. Alternatively Richards (1977) has shown that amino acids have an average cubic volume with an edge length of 5.3Å. In separate studies, Covell has used cubic lattices with sides of both lengths. He defined the allowed lattice movements as being of 5 types, all of which were strictly self avoiding.

His energy functions included an attractive term based on the potentials of mean force derived by Miyawaza and Jernigan (1985), and a surface exposure term. Simulated annealing or Monte Carlo algorithms were used for sampling conformations. In a series of small proteins ranging from 46 to 76 amino acids the average rms fit to the native folded state was 7.6 ±0.7Å

Another group who have pioneered the use of lattice models is that of Dill and co-workers at UCSF (Yue and Dill, 1995; Dill *et al.* 1995). In their work the cubic lattice typically does not have meaningful dimensions. Amino acids are classified as being of two types only, hydrophobic (H) or polar (P), and knowledge of the actual amino acid identity is ignored. The driving force for protein folding is the "hydrophobic collapse", i.e. the initial formation of a hydrophobic core. It has been proven in fact that a fold with an H core and the minimal surface area corresponds to the native structure (Dill *et al.* 1995). This of course applies only to globular proteins. The energy of the system is simply given by the total number of HH contacts. To fold the sequence, use is made of the constrained hydrophobic core construction method. This algorithm was developed to maximise the initial formation of a hydrophobic core, placing single "P"s on the outer edges of the grid. Sampling with this algorithm has been shown to be up to 10^{37} times faster than full atomistic models. Using these methods they have been able to find global minima for small proteins up to 88 amino acids including 4 helix bundles and α/β barrels (Yue and Dill, 1995).

The main problems with cubic lattice models are that they lose secondary structure resolution, they have an unrealistic geometric relationship to real proteins, they ignore the question of chirality and, until recently, the chain lengths simulated are too short to represent real proteins. Some of these problems have been overcome in an alternative lattice model developed by Skolnick and Kolinski (1990). A much more detailed cubic lattice of unit size is constructed which has some implied native structure. Both $C\alpha$ and $C\beta$ carbons are included at lattice points with the $C\alpha$s placed at (2,1,0) apart (i.e. similar to a knight's move in chess), thus giving inter-residue directionality. The $C\beta$ atoms are included with correct chirality. The inter-residue energy term is based on a Miyazawa and Jernigan hydrophobic scale (Miyawaza and Jernigan, 1985) applied to each $C\beta$. The surrounding 6 points of the $C\alpha$ are filled to give thickness to the bonds, hence giving realistic hard core repulsion, and hydrogen bond interactions are included. Using a Monte Carlo sampling, French bean plastocyanin was folded and found to produce a family of low energy conformations topologically identical to the native structure. This is much slower than other lattice models but, as implemented in the *Monsster* program (Skolnick *et al.* 1997) in combination with some other techniques such as the U-turn algorithm of Kolinski and co-workers (1997), it has been successful with many predictions.

GLOBULAR VS MEMBRANE PROTEINS

The discussion so far has concentrated on the structure prediction of soluble globular proteins, the class for which almost all the experimental structure determination has been performed. Another broad class however, which has not been discussed so far, is the water insoluble, membrane bound proteins. With the exception of the cryoelectron diffraction studies mentioned earlier, direct structural information is not readily available. This is due not only to the lack of solubility, but also to low expression levels, and to the fact that attempts to solubilise the proteins can cause denaturation or refolding. Structural information has mainly come from indirect evidence such as ligand binding SAR, site directed mutagenesis (SDM) (Savarese and Fraser, 1992) including the elegant cysteine scan method (Akabas et al. 1992; Javitch et al. 1994), photoaffinity labeling (Findlay and Pappin, 1986; Donnelly et al. 1989) spin labelling (Altenbach et al. 1996), antibody generation (Wang et al. 1989) and a host of other techniques which have given useful but limited data on the nature of these proteins.

From a medicinal or biological viewpoint however, these proteins are vitally important. It has been estimated that up to 40% of all known drugs in use act on one family of these alone, *viz.* the G-protein coupled receptors (GPCRs). Other families such as ion channels (eg. 5HT$_3$, GABA, Na+, K+, Ca+, etc) and ligand transporters (eg. 5HT reuptake) are also very important as drug targets. For this reason an enormous effort worldwide has gone into the prediction of their structure and a number of *de novo* techniques have been developed to deal with the special properties of these proteins. Most of this work has concentrated on GPCRs because of their medicinal importance and, because it is for this class that the majority of experimental evidence has been gathered. The brief discussion below will deal with GPCRs only. However we (Doughty et al. 1995; Doughty et al. 1998) and others (Guy and Conti, 1990; Bogusz et al. 1992) have carried out structure predictions on ion channels and at least one group has also made a model of a reuptake protein, the glucose transporter (Osguthorpe, 1992).

Predictions of G protein-coupled receptor structure

GPCR modelling has been the subject of numerous papers and a number of excellent reviews have been published on the subject (Ballesteros and Weinstein, 1995; Humblet and Mirzadegan, 1992). It is universally accepted that GPCRs consist of seven transmembrane (TM) regions which, from their sequence length, and analogy with the bacterial membrane protein, bacteriorhodopsin (BR), are presumed to be α-helical. Although there is no homology between GPCRs and BR (which itself is not a GPCR), it is not suprising that the latter has been extensively used as a template for the construction of GPCR models (Findlay and Eliopoulos, 1990; Lewell, 1992; Livingstone et al. 1992; Pardo et al. 1992; Trumpp-Kallmeyer et al. 1991; Trumpp-Kallmeyer et al. 1992; Zhang and Weinstein, 1993). The TM regions of these proteins show a high degree of sequence similarity, particularly among subfamilies. Thus multiple sequence alignments and the technique of hydropathic analysis (Kyte and Dolittle, 1982; Engelman et al. 1986) have been used to determine the extent of the TM bundles. The early bovine rhodopsin density map (Schertler et al. 1993) and the very recent publication of slices of density through the membrane of frog rhodopsin (Unger et al. 1997) have shown that the 7 helices have a different packing from BR. The bovine rhodopsin paper led several groups to modify their earlier BR based models (Bourdon et al. 1997). However the observed differences also support the efforts that have gone into the development of *de novo* methods of packing the helices of GPCRs (Kontoyianni *et al.* 1996; Evardsen *et al.* 1991; Evardsen *et al.* 1992). In some early work, we used an algorithm based on maximization of the hydrophobic energy (Blaney, 1991; Blaney and Tennant, 1996a), calculated by the method of Eisenberg (Eisenberg and McLachlan, 1986). This unfortunately gave untilted helices with a very large TM bundle cavity (Blaney and Tennant, 1996a). A recognition of the fact that functionally important (often polar) residues were more highly conserved and faced towards the interior of the TM bundle, led to the development of the Fourier transform method of sequence conservation analysis. The periodicity of regular α-helices led Donnelly and co-workers (1989) to derive a conservation moment which defined the interior face of each helix. This is often opposite in direction to the hydrophobic moment which has also been used in GPCR modelling to define the membrane facing side (MaloneyHuss and Lybrand, 1992). We have extended this Fourier method to define a 2-D helical wheel technique, known as Helanal, which maps out not only the "Donnelly moment", but also defines a hydrophobic arc (Blaney and Tennant, 1996a; Blaney and Tennant, 1996b). This may be thought of as a measure of the exposure of each helix to the lipid bilayer. Using all 7 helanal wheels it is possible to manipulate these in "real-time" on the screen, giving a packing arrangement which maximises the hydrophobic arc exposure and ensures that the conservation moment is inwards. This 2-D representation can automatically then be converted to a 3-D model (see Figure 7).

Many GPCR template structures have been developed from these and other techniques. They have been incorporated into an automated program, GPCR_Builder (Blaney and Tennant, 1996a) which can now go from a sequence to a 3-D structure in a few minutes. Such models have been extensively used in our rational drug design work (Forbes *et al.* 1996; Blaney, 1996).

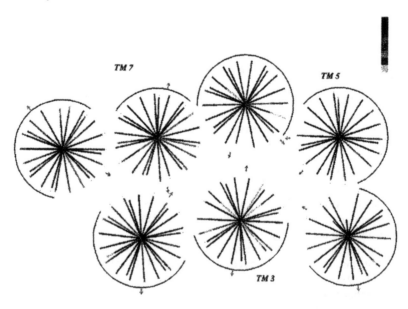

Figure 7a. The 2-D Helanal wheels are set up on the screen and are manipulated in "real time" so as to maximise the hydrophobic arcs.

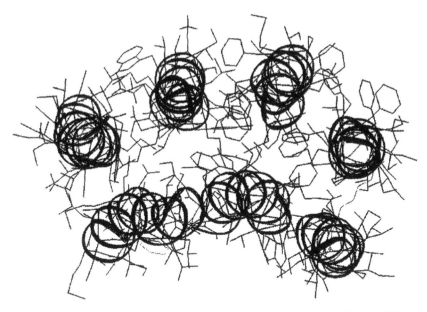

A Model of the 5HT2c Receptor from Helanal & Constrained MD–MM

Figure 7b. The seven Helanal wheels above are then automatically converted to a 3-D structure before subsequent minimisation.

CONCLUSIONS & COMMENTS

Much credit in the promotion of the field of protein folding prediction must be given to John Moult who issued a challenge to those working in the area. A number of target sequences, whose structure elucidations were nearing completion, were given out. Groups were invited to make predictions on these and then compare their results against the actual native structures. Two such meetings, known as CASP1 and CASP2, have been held (in 1994 and 1996) at Asilomar, California and the results have been published (CASP1, 1995). They have shown that the protein folding problem is certainly by no means solved but significant advances have been and continue to be made. Together with the increasing speed of computers, it is conceivable that it will no longer be an unattainable goal. "Black Box" programs are becoming available which do a good job on many globular proteins up to 100 amino acids, eg. the MONSSTER program of Skolnick and Kolinski (Skolnick *et al.* 1997). It has already been noted that the number of folds may indeed be finite and, as suggested by Paul Bash (1997), structural analysis of a subset of novel proteins arising from the genome projects, carefully selected so as to maximise dissimilarity with known structures, may eventually give us a complete set of folds. The next decade certainly promises to be an exciting time for the field of *de novo* protein structure prediction.

REFERENCES

Abagyan, R., Totrov, M. and Kuznetsov, D. (1994) ICM - a new method for protein modeling and design: applications to docking and structure prediction from the distorted native conformation. *Journal of Computational Chemistry*, **15**, 488-506.

Akabas, M.H., Stauffer, D.A., Xu, M. and Karlin, A. (1992) Acetylcholine receptor channel structure probed in cysteine-substitution mutants. *Science*, **258**, 307-310.

Altenbach, C., Yang, K., Farrens, D.L., Khorana, H. G. and Hubbell, W.L. (1996) Structural Features and Light-Dependent Changes in the Cytoplasmic Interhelical E-F Loop Region of Rhodopsin: A Site-Directed Spin-Labeling Study. *Biochemistry*, **35**, 12470-12478.

Altschul, S.F., Madden, T.L., Schaffer, A.A., Zhang, J., Zhang, Z., Miller, W. and Lipman, D.J. (1997) Gapped BLAST and PSI-BLAST: a new generation of protein database search programs, *Nucleic Acids Research*, **25**, 3389-3402.

Anfinsen, C.B. (1973) Principles that govern the folding of protein chains. *Science*, **181**, 223-230.

Ballesteros, J.A. and Weinstein, H. (1995) Integrated methods for the construction of three-dimensional models and computational probing of structure-function relations in G-protein coupled receptors. *Methods in Neurosciences*, **25**, 366-427

Bash, P. (1997) Personal communication

Bernstein, F.C., Koetzle, T.F., Williams, G.J.B., Meyer, E.F., Brice, M.D., Rodgers, J.R., Kennard, O., Shimanouchi, T. and Tasumi, M. (1977) The protein data bank: a computer-based archival file for macromolecular structures. *Journal of Molecular Biology*, **112**, 535-542.

Blaney, F.E. (1991) A Unified Model for Serotonin-5HT2 Antagonist Receptor Binding. Paper presented at the *"International Society of Quantum Biology and Pharmacology (ISQBP) Presidential Meeting"*, Stanford, California, USA.

Blaney, F.E. (1996) The use of GPCR models in the design of selective 5HT2C and 5HT1B antagonists. Paper presented at *"G Protein-Coupled Receptors: Structure & Function"* International meeting held in honour of Dr Paul Janssen, Beerse, Belgium.

Blaney, F.E. and Tennant, M. (1996a) Computational tools and results in the construction of G protein-coupled receptor models. In: *Membrane Protein Models*, J.B.C. Findlay (Ed.), Oxford, Bios Scientific Publishers Ltd. pp 161-176.

Blaney, F.E. and Tennant, M. (1996b) Helanal - A Novel Algorithm for the Construction of Transmembrane Helical Bundles. Poster presented at the *"International Conference on Protein Folding and Design"*, Bethesda, MD., USA.

Blundell, T.L., Elliott, G., Gardner, S.P., Hubbard, T., Islam, S., Johnson, M., Mantafounis, D., Murray-Rust, P. and Overington, J. (1989) Protein engineering and design. *Philosophical Transactions of the Royal Society, London, B*, 324, 447-460.

Bogusz, S., Boxer, A. and Busath, D.D. (1992) A SS1-SS2 β-barrel structure for the voltage-activated potassium channel. *Protein Engineering*, **5**, 285-293.

Boscott, P.E., Barton, G.J. and Richards, W.G. (1993) Secondary structure prediction for modeling by homology. *Protein Engineering*, **6**, 261-266.

Bourdon, H., Trumpp-Kallmeyer, S., Schreuder, H., Hoflack, J., Hibert, M. and Wermuth, C. (1997) Modeling of the binding site of the human m1 muscarinic receptor: experimental validation and refinement. *Journal of Computer-Aided Molecular Design*, **11**, 317-332.

Bowie, J.U., Lüthy, R. and Eisenberg, D. (1991) A method to identify protein sequences that fold into a known three-dimensional structure. *Science*, **253**, 164-170.

Brandon, C. and Tooze, J. (1991) *Introduction to Protein Structure*. New York and London: Garland Publishing.

Brooks, B.R., Bruccoleri, R.E., Olafson, B.D., States, D.J., Swaminathan, S. and Karplus, M. (1983) CHARMM. A program for macromolecular energy, minimisation and dynamics calculations. *Journal of Computational Chemistry*, **4**, 187-217.

Browman, M.J., Burgess, A.W., Dunfield, L.G. , Rumsey, S.M., Endres, G.F. and Scheraga, H.A. (1977) United Atom Conformational Energy Program for Peptides (UNICEPP). Quantum Chemistry Program Exchange. QCPE #361, Indiana university, Chemistry Dept., Bloomington.

Browman, M.J., Carruthers, L.M., Kashuba, K.L., Momany, F.A., Pottle, M.S., Rosen, S.P. , Rumsey, S.M., Endres, G.F. and Scheraga, H.A. (1975) Empirical Conformational Energy Program for Peptides (ECEPP). Quantum Chemistry Program Exchange. QCPE #286, Indiana university, Chemistry Dept., Bloomington.

Bryant, S.H. and Lawrence, C.E. (1993) An empirical energy function for threading protein sequence through folding motif. *Proteins: Structure, Function, Genetics*, **16**, 92-112.

CASP1 Results (1995) *Proteins: Structure, Function, Genetics.* **23**, Number 3.

Chou, P.Y. and Fasman, G.D. (1974) Prediction of protein conformation. *Biochemistry*, **13**, 222-245.

Covell, D.G. (1992) Folding protein α-carbon chains into compact forms by Monte Carlo methods. *Proteins: Structure, Function, Genetics*, **14**, 409-420.

Covell, D.G. (1994) Lattice model simulations of polypeptide chain folding. *Journal of Molecular Biology*, **235**, 1032-1043.

Crippen, G.M. (1991) Prediction of protein folding from amino acid sequence over discrete conformation spaces. *Biochemistry*, **30**, 4232-4237

Crippen, G.M. and Havel, T.F. (1988) *Distance Geometry and Molecular Conformation*. Taunton: Research Studies Press.

Curtis, B.M., Presnell, S.R., Srinivasan, S., Sassenfeld, H., Klinke, R., Jeffrey, E., Cosman, D., March, C.J. and Cohen, F.E. (1991) Experimental and theoretical studies of the three-dimensional structure of human interleukin-4. *Proteins: Structure, Function, Genetics*, **11**, 111-119.

Dayhoff, M.O., Schwartz, R.M. and Orcutt, B.C. (1972). *Atlas of protein sequence and structure.* **5(S3)**, 345.

Dill, K.A., Bromberg, S., Yue, K., Fiebig, K.M., Yee, D.P., Thomas, P.D. and Chan, H.S. (1995) Principles of protein folding - a perspective from simple exact models. *Protein Science*, **4**, 561-602.

Donnelly, D., Johnson, M.S.,. Blundell, T.L. and Saunders, J. (1989) An analysis of the periodicity of conserved residues in sequence alignments of G protein-coupled receptors - Implications for the 3 dimensional structure. *FEBS Letters,* **251**, 109-116.

Doughty, S.W., Blaney, F.E., Orlek, B.S. and Richards, W.G. (1998) A molecular mechanism for toxin block in N-type calcium channels. *Protein Engineering*, **11**, in press.

Doughty, S.W., Blaney, F.E. and Richards, W.G. (1995) Models of ion pores in N-type voltage-gated calcium channels. *Journal of Molecular Graphics*, **13**, 342-8.

Editorial (1997) Structure and the genome. *Nature Structural Biology*, **4**, 329-330

Eisenberg, D. and McLachlan, A.D. (1986) Solvation energy in protein folding and binding. *Nature,* **319**, 199-203.

Engelman, D.M., Steitz, T.A. and Goldman, A. (1986) Identifying non-polar transbilayer helices in amino-acid sequences of membrane proteins. *Annual Reviews in Biophysics & Biophysical Chemistry,* **15**, 321-353.

Evans, J.N.S. (1995) *Biomolecular NMR Spectroscopy.* Oxford, New York and Tokyo: Oxford University Press.

Evardsen, O., Sylte, I. and Dahl, S. (1991) Molecular dynamics of dopamine at the D_2 receptor. *Proceedings of the National Academy of Science, USA,* **88,** 8111-8115.

Evardsen, O., Sylte, I. and Dahl, S. (1992) Molecular dynamics of serotonin and ritanserin interacting with the 5-HT_2 receptor. *Molecular Brain Research* **14,** 166-178.

Feng, D.F. and Doolittle, R.F. (1987) Progressive sequence alignment as a prerequisite to correct phylogenetic trees. *Journal of Molecular Evolution*, **25**, 351-360.

Findlay, J.B.C. and Pappin, D.J.C. The opsin family of proteins. (1986) *Biochemical Journal*, **238**, 625-642,

Findlay, J. and Eliopoulos, E. Three-dimensional modelling of G protein-linked receptors. (1990) *Trends in Pharmacological Sciences,* **11**, 492-499.

Finkelstein, A.V. and Reva, B.A. (1991) A search for the most stable folds of protein chains. *Nature*, **351**, 497-499.

Forbes, I.T., Dabbs, S., Duckworth, D.M., Ham, P., Jones, G.E., King, F.D., Saunders, D.V., Blaney, F.E., Naylor, C.B., Baxter, G.S., Blackburn, T.P., Kennett, G.A. and Wood, M.D. (1996) Synthesis, Biological Activity, and Molecular Modeling Studies of Selective 5-HT2C/2B Receptor Antagonists. *Journal of Medicinal Chemistry*, **39**, 4966-4977.

Garnier, J., Osguthorpe, D.J. and Robson, B. (1978) Analysis of the accuracy and implications of simple methods for predicting the secondary structure of globular proteins. *Journal of Molecular Biology*, **120**, 97-120.

Godzik, A., Kolinski, A. and Skolnick, J. (1993) De novo and inverse folding predictions of protein structure and dynamics. *Journal of Computer Aided Molecular Design*, **7**, 397-438.

Gregoret, L.M. and Cohen, F.E. (1990) Novel method for the rapid evaluation of packing in protein structures. *Journal of Molecular Biology*, **211**, 959-74.

Gunn, J.R., Monge, A., Friesner, R.A. and Marshall, C.H. (1994) Hierarchical algorithm for computer modelling of protein tertiary structure: folding of Myoglobin to 6.2Å resolution. *Journal of Physical Chemistry*, **98**, 702-711.

Guy, H.R. and Conti, F. (1990) Pursuing the structure and function of voltage-gated channels. *Trends in Neurosciences*, **13**, 201-206.

Havel, T.F. (1993) Predicting the structure of the flavodoxin from Eschericia coli by homology modeling, distance geometry and molecular dynamics. *Molecular Simulation*, **10**, 175-210.

Hemmings, A., Foundling, S.I., Sibanda, B.L., Wood, S.P., Pearl, L.H. and Blundell, T.L. (1985) Energy calculations on aspartic proteinases: human renin, endothiapepsin and its complex with an angiotensinogen fragment analog, H-142. *Biochemical Society Transactions*, **13**, 1036-1041.

Henderson, R., Baldwin, J.M., Ceska, T.A., Zemlin, F., Beckmann, E. and Downing, K.H. (1990) Model for the structure of bacteriorhodopsin based on high-resolution electron cryo-microscopy. *Journal of Molecular Biology*, **213**, 899-929.

Humblet, C. and Mirzadegan, T. (1992) Three-dimensional models of G-protein coupled receptors. *Annual Reports in Medicinal Chemistry*, **27**, 291-300.

INSIGHT. Molecular Simulations Inc., 9685 Scranton Road, San Diego, CA, 92121-3752.

Javitch, J.A., Li, X., Kaback, J. and Karlin, A. (1994) A cysteine residue in the third membrane-spanning segment of the human D2 dopamine receptor is exposed in the binding-site crevice. *Proceedings of the National Academy of Science, USA*, **91**, 10355-10359.

Jones, D.T., Taylor, W.R. and Thornton, J.M. (1992) A new approach to protein fold recognition. *Nature*, **358**, 86-89.

Kabsch, W. and Sander, C. (1983) How good are predictions of protein secondary structure? *FEBS Letters*, **155**, 179-182.

Kabsch, W. and Sander, C. (1984) On the use of sequence homologies to predict protein structure: Identical pentapeptides can have completely different conformations. *Proceedings of the National Academy of Science, USA*, **81**, 1075-1078.

Kolinski, A., Skolnick, J., Godzik, A. and Hu, W. (1997) A method for the prediction of surface "U"-turns and transglobular connections in small proteins. *Proteins: Structure, Function, Genetics*, **27**, 290-308.

Kontoyianni, M., DeWeese, C., Penzotti, J.E. and Lybrand, T.P. (1996) Three-dimensional models for agonist and antagonist complexes with β2 adrenergic receptor. *Journal of Medicinal Chemistry*, **39**, 4406-4420.

Kostrowicki, J. and Scheraga, H.A. (1992) Application of the diffusion-equation method for global optimization to oligopeptides. *Journal of Physical Chemistry*, **96**, 7442-7449.

Kyte, J. and Dolittle, R.F. (1982) A simple method for displaying the hydropathic character of a protein. *Journal of Molecular Biology* **157**, 105-132.

Laskowski, R.A., MacArthur, M.W., Moss, D.S. and Thornton, J. M. (1993) PROCHECK: a program to check the stereochemical quality of protein structures. *Journal of Applied Crystallography*, **26**, 283-91.

Levinthal, C. (1968) Are these pathways for protein folding? *Journal of Chemical Physics*, **65**, 44-5.

Lewell, X.Q. (1992) A model of the adrenergic beta-2 receptor and binding sites for agonist and antagonist. *Drug Design Discov*ery, **9**, 29-48.

Lim, V.I. (1974) Algorithms for prediction of α-helical and β-structural regions in globular proteins. *Journal of Molecular Biology*, **88**, 873-94.

Livingstone, C.D., Strange, P.G. and Naylor, L.H. (1992) Molecular modeling of D2-like dopamine receptors. *Biochemical Journal*, **287**, 277-282.

Liwo, A., Pincus, M.R., Wawak, R.J., Rackovsky, S. and Scheraga, H.A. (1993) Prediction of protein conformation on the basis of a search for compact structures: test on avian pancreatic polypeptide. *Protein Science*, **2**, 1715-1731.

Lozano, J.J., López-de-Briñas, E., Centeno, N.B., Guigó, R. and Sanz, F. (1997) Three-dimensional modelling of human cytochrome P450 1A2 and its interaction with caffeine and MeIQ. *Journal of Computer-Aided Molecular Design*. **11**, 395-408.

Maiorov, V.N. and Crippen, G.M. (1992) Contact potential that recognizes the correct folding of globular proteins. *Journal of Molecular Biology,* **227**, 876-888.

MaloneyHuss, K. and Lybrand, T.P. (1992) Three-dimensional structure for the β_2 adrenergic receptor protein based on computer modeling studies. *Journal of Molecular Biology,* **225**, 859-871.

Martin, A.C.R., Cheetham, J.C. and Rees, A.R. (1989) Modeling antibody hypervariable loops: a combined algorithm. *Proceedings of the National Academy of Science, USA,* **86**, 9268-9272.

Miyawaza, S. and Jernigan, R.L. (1985) Estimation of effective interresidue contact energies from protein crystal structures; quasi-chemical approximation. *Macromolecules,* **18**, 534-552.

Morris, G.M. (1988) The matching of protein sequences using color intrasequence homology displays. *Journal of Molecular Graphics,* **6**, 135-140.

Moult, J. and James, M.N.G. (1986) An algorithm for determining the conformation of polypeptide segments in proteins by systematic search. *Proteins: Structure, Function, Genetics,* **1**, 146-163.

Murzin, A.G., Brenner, S.E., Hubbard, T. and Chothia, C. (1995) SCOP: a structural classification of proteins database for the investigation of sequences and structures. *Journal of Molecular Biology,* **247**, 536-540.

Neuhaus, D. and Williamson, M. (1989*) The Nuclear Overhauser Effect in Structural and Conformational Analysis.* New York: VCH Publishers.

Orengo, C. A., Michie, A. D., Jones, S., Jones, D. T., Swindells, M. B. and Thornton, J. M. (1997) CATH - a hierarchic classification of protein domain structures. *Structure,* **5**, 1093-1108.

Osguthorpe, D. (1992) personal communication.

Osguthorpe, D. (1997) personal communication.

Ouzounis, C., Sander, C. Scharf, M. and Schneider, R. (1993) Prediction of protein structure by evaluation of sequence-structure fitness: Aligning sequences to contact profiles derived from 3D structures. *Journal of Molecular Biology,* **232**, 805-825.

Pardo, L., Ballesteros, J.A. Osman, R. and Weinstein, H. (1992) On the use of transmembrane domain of bacteriorhodopsin as a template for modeling the three-dimensional structure of guanine nucleotide-binding regulatory protein-coupled receptors. *Proceedings of the National Academy of Science, USA,* **89**, 4009-4012.

Pearl, L.H. and Taylor, W.R. (1987) A structural model for the retroviral proteases. *Nature,* **329**, 351-354.

Peitsch, M.C. (1996) ProMod and Swiss-Model: Internet-based tools for automated comparative protein modelling. *Biochemical Society Transactions,* **24**, 274-279.

Ponder, J.W. and Richards, F.M. (1987) Tertiary templates for proteins. *Journal of Molecular Biology*, **193**, 775-791.

QUANTA. Molecular Simulations Inc., 9685 Scranton Road, San Diego, CA, 92121-3752.

Ramachandran, G.N. and Sassiekharan, V. (1968) Conformation of polypeptides and proteins. *Advances in Protein Chemistry*, **28**, 283-437.

Richards, F.M. (1977) Areas, volumes, packing, and protein structure. *Annual Reviews of Biophysics & Bioengineering*, **6**, 151-176.

Rost, B. and Sander, C. (1993a) Improved prediction of protein secondary structure by use of sequence profiles and neural networks. *Proceedings of the National Academy of Science, USA*, **90**, 7558-7562.

Rost, B. and Sander, C. (1993b) Prediction of protein secondary structure at better than 70% accuracy. *Journal of Molecular Biology*, **232**, 584-599.

Sali, A. and Blundell, T.L. (1993) Comparative protein modeling by satisfaction of spatial restraints. *Journal of Molecular Biology*, **234**, 779-815.

Sali, A., Overington, J.P., Johnson, M.S. and Blundell, T.L. (1990) From comparisons of protein sequences and structures to protein modeling and design. *Trends in Biochemical Science*, **15**, 235-240.

Savarese, T.M. and Fraser, C.M. (1992) In vitro mutagenesis and the search for structure-function relationships among G protein-coupled receptors. *Biochemical Journal*, **283**, 1-19.

Schertler, G.F.X., Villa, C. and Henderson, R. (1993) Projection structure of rhodopsin. *Nature*, **362**, 770-772.

Sippl, M. (1990) Calculation of conformational ensembles from potentials of mean force. An approach to knowledge-based prediction of local structures in globular proteins. *Journal of Molecular Biology*, **213**, 859-883.

Sippl, M.J. and Weitckus, S. (1992) Detection of native-like models for amino acid sequences of unknown three-dimensional structure in a database of known protein conformations. *Proteins: Structure, Function, Genetics*, **13**, 258-271.

Skolnick, J., Kolinski, A. and Ortiz, A.R. (1997) MONSSTER: a method for folding globular proteins with a small number of distance restraints. *Journal of Molecular Biology*, **265**, 217-241.

Skolnick, J. and Kolinski, A. (1990) Simulations of the folding of a globular protein. *Science*, **250**, 1121-1125.

Smellie, A., Teig, S.L. and Towbin, P. (1995) Poling: promoting conformational variation. *Journal of Computational Chemistry*, **16**, 171-187.

Summers, N.L., Carlson, W.D. and Karplus, M. (1987) Analysis of side-chain orientations in homologous proteins. *Journal of Molecular Biology*, **196**, 175-198.

SYBYL, 6.3 (1996). Tripos, Inc., 1699 S. Hanley Road, St.Louis, MO., 63144-2913.

Thompson, J.D., Higgins, D.G. and Gibson, T.J. (1994) CLUSTAL W: improving the sensitivity of progressive multiple sequence alignment through sequence weighting, position-specific gap penalties and weight matrix choice. *Nucleic Acids Research*, **22**, 4673-4680.

Trumpp-Kallmeyer, S., Hibert, M.F., Bruinvels, A. and Hoklack, J. (1991) Three-dimensional models of neurotransmitter G-binding protein-coupled receptors. *Molecular Pharmacology* **40**, 8-15.

Trumpp-Kallmeyer, S., Hoklack, J., Bruinvels, A. and Hibert, M.F. (1992) Modeling of G-protein-coupled receptors: application to dopamine, adrenaline, serotonin, acetylcholine and mammalian opsin receptors. *Journal of Medicinal Chemistry* **35**, 3448-3462.

Unger, V.M., Hargrave, P.A., Baldwin, J.M. and Schertler, G.F.X. (1997) Arrangement of rhodopsin transmembrane α-helixes. *Nature*, **389**, 203-206.

Vriend, G., Sander, C. and Stouten, P.F.W. (1994) A novel search method for protein sequence-structure relations using property profiles. *Protein Engineering*, **7**, 23-29.

Wang, H.Y., Lipfert, L., Malbon, C.C. and Bahouth, S. (1989) Site-directed anti-peptide antibodies define the topography of the β-adrenergic receptor. *Journal of Biological Chemistry*, **264**, 14424-14431.

Weiner, S.J., Kollman, P.A., Case, D.A., Singh, U.C., Ghio, C., Alagona, G., Profeta, S. and Weiner, P. (1984) A new force field for molecular mechanical simulation of nucleic acids and proteins. *Journal of the American Chemical Society*, **106**, 765-784.

Wilmot, C.M. and Thornton, J.M. (1988) Analysis and prediction of the different types of β-turn in proteins. *Journal of Molecular Biology*, **203**, 221-232.

Yue, K. and Dill, K.A. (1995) Forces of tertiary structural organization in globular proteins. *Proceedings of the National Academy of Science, USA*, **92**, 146-50.

Yue, K. and Dill, K.A. (1996) Folding proteins with a simple energy function and extensive conformational searching. *Protein Science*, **5**, 254-261.

Zhang, D. and Weinstein, H. (1993) Signal transduction by a 5-HT$_2$ receptor: a mechanistic hypothesis from molecular dynamics simulations of the three-dimensional model of the receptor complexed to ligands. *Journal of Medicinal Chemistry*, **36**, 934-938.

Zuiderweg, E.R.P., Scheek, R.M., Boelens, R., van Gunsteren, W.F. and Kaptein, R. (1985) Determination of protein structures from nuclear magnetic reference data using a restrained molecular dynamics approach: the Lac Repressor DNA binding domain. *Biochimie*, **67**, 707-715.

LIMITATIONS OF MOLECULAR BIOLOGY IN PESTICIDE DISCOVERY

ALAN AKERS

BASF Agricultural Research Station, Limburgerhof, Germany.

> *"Life is understood backwards, but must be lived forwards."*
> Søren Kierkegaard

ABSTRACT

Molecular biology and genomics are powerful tools for the *post hoc* rationalisation in molecular terms of the biological activity of compounds which emerge from screening programmes. They are not predictive tools and cannot replace random screening in pesticide discovery. Had such methods been used in the past, several currently successful pesticides would probably never have been developed. They generate many more hypotheses than can be tested biochemically, which are neither more nor less likely to be true than any other form of speculation. To begin with molecules and expect to achieve a predetermined biological result is not only greatly to underestimate the complexity of biological systems, but an attempt to reverse the process of scientific understanding in which problems are solved empirically and understood retrospectively.

INTRODUCTION

Random screening remains the most successful strategy for pesticide discovery. So successful has it been, that the discovery of new compounds which offer any advantages over the extremely effective ones already discovered becomes increasingly more difficult and costly. A point of vanishing returns could appear to threaten this strategy when ever greater numbers of compounds must be synthesised and screened at very high cost to achieve ever diminishing rates of success. Molecular biology seems to offer a way out of this dilemma by reducing or even eliminating the random component in the discovery process and replacing it with a programme of rational design which exploits the recent advances in molecular genetics and biotechnology.

Considering the vast investments of material and intellectual resources in molecular biology and genomics, it would be very surprising if they did not contribute to understanding the molecular aspects of pesticide action. However, when new approaches bring progress to problems that have resisted previous methods, there is sometimes more than a small element of hubris and hyperbole in the claims made for what they can achieve. If every claim for the power and potential of molecular genetics were to be believed, all the apparent contingencies in nature are, in reality, no more than temporary gaps in our knowledge which will constantly diminish as genetic data banks become more complete and computer programmes designed to extract and interpret the information contained in genomes become more sophisticated. An intellectual climate so dominated by the single concept that there is a gene for everything readily fosters the impression that molecular genetics offers a direct route to pesticide discovery which renders obsolete the more traditional methods of inquiry.

The *reductio ad absurdium* of such a viewpoint is that pesticide discovery could eventually become largely a matter of predictive virtual biology and the vulnerabilities of such weeds, pests and diseases as manage to persist, could be deduced from their genomes. Appropriate protective measures could then be designed from first principles. Even the most fervent apostles of molecular biology, cloistered from any contact with the practical rigours of applied science, may shrink from advancing such a proposal. Nevertheless, startling claims continue to be made. One of the more modest is that sequencing the yeast genome ultimately offers "...the complete understanding of how a eukaryotic cell functions." (Johnston, 1996) It is hardly surprising, therefore, that there have been great expectations for what molecular genetics might contribute to solving the problems of plant protection. As far as pesticide discovery is concerned, such expectations have yet to be realised and by far the great majority of compounds currently on the market, including those most recently introduced, originated from random screening programmes. Although the leads times required to bring plant protection products to the market are admittedly long, molecular biology can no longer claim to be a nascent technology. Perhaps the reasons it has yet to bring any short cuts to pesticide discovery and has not replaced random screening, are that agricultural systems are not entirely genetically determined and, even those aspects that are, cannot be predicted from the base sequences of DNA.

The yields of agricultural crops are the products of immensely complex interactions between biological, geophysical and meteorological systems, modified by husbandry practices which have been empirically optimised, literally over millennia. The relatively recent introduction of agrochemicals provided powerful additional tools for further empirical optimisation of the system and resulted in great yield improvements. The intended, but not always the exclusive influence of pesticides, is to inhibit the capacity of weeds, pests and diseases to compete for resources. In some cases, causal links have been established between the effects of pesticides on the functions of specific proteins and the observed macroscopic biological consequences. This led to the concept of targets and rational design which, in one of its earliest incarnations, reasoned that, since pesticides interact with proteins and alter their function, biochemical assays of protein function could be used to discover and optimise new lead structures, once target proteins had been identified. Such studies have contributed to understanding pesticide activity, but biochemists would require a faith in their discipline amounting to myopia if they failed to concede that enzyme inhibition data as tools for predicting useful activity at higher levels of biological complexity are, in practical terms, of limited value (Akers *et al.,* 1991). They provide a salutary reminder that, in the complexity of biological systems, "target" is more of a conceptual convenience than a biological reality.

It is, therefore, not unreasonable also to question how molecular biologists expect to slice through the Gordian knot of biological complexity and achieve a predetermined aim by beginning at an even lower, in fact, at the lowest possible level of biological complexity with DNA. That it can contribute to pesticide discovery is not disputed, but molecular biology can only help as biochemistry helps, as an additional *post hoc* aid to understanding, not as an *a priori* predictive tool that can revolutionise the process. An awareness of their specific limitations is essential to ensure that the powerful tools of molecular biology are employed where they are most likely to be productive.

LIMITATIONS OF THE MOLECULAR APPROACH

Target Selection is Random, not Rational

No system of genomic scrutiny *per se* can identify gene sequences encoding new target proteins whose inhibition would achieve predetermined results; this would be equivalent to scanning the "zeroes" and "ones" of the binary code of a computer programme and expecting to discover, not only which parts of it encode word processing capacity, but also what was likely to be written by the persons using it. Only the most determinedly virtual of virtual biologists would make such a naïve assertion, but there are claims that molecular biology can identify and validate biochemical targets by techniques such as antisense and gene deletion.

The first and perhaps the greatest problem with this approach is that of target selection. No particular molecular biological expertise is required to predict that inhibitors of photosynthesis may have herbicidal activity, or that inhibitors of nervous transmission may be toxic to insects, but these areas are largely covered, if not saturated. The challenge is to find new targets. Here genomics has been very useful in revealing the scale of the problem, if not in solving it. The yeast genome probably encodes around 6000 proteins and may be one of the smaller eukaryotic genomes; the organisation and life cycles of more complex, multicellular organisms are likely require many more genes. The time and costs involved in a single antisense or gene deletion experiment preclude the systematic investigation of whole genomes as a viable strategy of target selection. Taking into account that the relatively small increase in complexity when transferring from greenhouse tests to field trials produces inherent unpredictability, it is clear that there can be no *a priori* reasons for selecting any specific gene as being likely to encode a protein whose inhibition could conceivably control populations of target organisms under field conditions. If there were, pesticide discovery would be considerably easier than it is. Considering the odds against success represented by the tens of thousands of pleiotropic genes in target organisms and the number of times knockouts of seemingly vital functions are reported without any observable effects on the phenotype, it must be conceded that target selection for antisense or gene deletion experiments is, like pesticide discovery itself, essentially a matter of random screening at the molecular genetic level, and with no better guarantee of success.

Antisense and Gene Deletion do not Validate Novel Targets

The quality of the information that antisense and gene deletion experiments can deliver should also be considered. A legitimate way of testing their validity is to examine reports where genes encoding the known target enzymes of successful products have been deleted, and appraise whether or not such experiments would have identified those targets. The sterol biosynthetic pathway is a major target for agricultural fungicides and three of the known target enzymes within it have been deleted in bakers' yeast. Deletion of the *erg 11* gene which encodes sterol 14α-demethylase, the main target of the azole fungicides, initially resulted in cells which would only grow anaerobically in medium supplemented with ergosterol (Loper, 1992). However, after only two days, spores containing disrupted copies of the gene began to grow aerobically on non-supplemented agar. Sterol analysis showed that they contained only 14α-methyl sterols, so there had been no reverse mutation of *erg 11*. An additional, spontaneous mutation which allowed aerobic growth in the presence of 14α-methyl sterols had occurred. In *Candida albicans* the additional mutation was not necessary and cultures with a deleted *erg 11* gene grew immediately under aerobic conditions (Bard *et al.*, 1993). If such experiments had been the sole criteria to assess the suitability of the sterol 14α-demethylase as a target, one of the most successful groups of fungicides may never have been developed.

Slightly less ambiguous indications of possible target status were given in one report of the deletion of the sterol Δ^{14}-reductase gene, which encodes one of the known targets of morpholine fungicides (Baloch *et al.*, 1984). The disrupted strain would not grow aerobically, even in the presence of exogenous sterols, but an additional mutation which allowed very low levels of ergosterol uptake permitted aerobic growth (Marcireau *et al.*, 1992). This is certainly a very positive indication. However, in other reports, mutants of yeast (Ladevèze *et al.*, 1993) and *Neurospora crassa* (Ellis *et al.*, 1991) were aerobically viable without exogenous sterols, in spite of being deficient in Δ^{14}-reductase activity. Disruption of the *erg 2* gene encoding the other know target of the morpholines, the sterol Δ^{8-7}-isomerase, had no effect on viability (Ashman *et al.*, 1991). Thus, in only some cases, and then only in the context of a great deal of other biochemical and physiological information, gene disruption experiments may provide supplementary, but not definitive, indications that a particular enzyme could be a target for fungicides. This may be useful information, but it hardly merits the air of certainty implied by the term "target validation."

The Gap Between Molecular Genetics and Biochemistry

Perhaps a truly rational strategy, or at least a pragmatic one, would be simply to scan the literature for reports of deletion or antisense experiments which have already identified specific genes as being essential for viability. Unfortunately, although many genes have been investigated, relatively few have been assigned definite biochemical functions. The widening gap between molecular genetic data and the corresponding biochemistry is the greatest challenge in current molecular biological research and the greatest obstacle to commercial exploitation of genomics for the discovery of new, biologically active compounds. A gene encoding a product without a proven biochemical function is of little value to plant protection research, even if it is known to be essential.

Since, in the molecular biological paradigm, protein function is a property of structure and structure, at least in part, is determined by amino acid sequence, to what extent is it possible to predict protein functions from amino acid sequences or, allowing for introns, editing and splicing, from the base sequences of genes? The most successful method of doing this is not actually predictive, but merely comparative. If the sequence of a newly identified gene is very similar to that of a gene encoding a protein of known biochemical function, this is extremely good evidence that the protein encoded by the newly identified gene is likely to have the same function, bearing in mind that an unknown proportion of proteins may have more than one function. Biochemical confirmation, which may not always be a trivial matter, is required before the assigned function can be accepted without reservation, but sequence comparison is the most powerful tool for relating sequence to function and its value will increase as the functions of more proteins are discovered. However, the primary knowledge of the functions of proteins on which the success of sequence comparison depends does not originate in molecular biology. Its source is the long history of essentially chance discoveries during biochemical investigations and this remains the rate-limiting step in the discovery of new protein functions.

The latter is likely to remain a slow and uncertain process, which raises the question of how much significance, in terms of function, can be assigned to lower levels of sequence homology. There are "sequence motifs" which suggest that a protein may bind, for example, calcium or NADH, but they do not prove that it actually does so or indicate how such putative binding may be indicative of a specific biochemical function. Sequence homology may suggest that a newly discovered gene encodes a protein belonging to a group or family with related functions, such as protein kinases or cytochrome P-450s. This is certainly interesting and may suggest a fruitful line of investigation, but it leaves many possibilities open. Protein kinases, for example, modulate the activity of many enzymes, but to be of practical value in the search for new pesticide targets it would be necessary to know which enzymes are their substrates. Again, the molecular biological approach does more to reveal the size of the problem than to solve it. The peptide sequences recognised by protein kinases are between 9 and 12 amino acids long. With the possibility of 20 different amino acids at each position, there are between 20^9 and 20^{12} different possible substrates. A systematic search, even with the most advanced high throughput equipment, may thus prove to be less than practical. The importance of cytochrome P-450 monooxygenases in pesticide biochemistry, both as targets in themselves and as the mediators of resistance to insecticides and the selectivity of herbicides, has been well established by biochemical investigations. At first sight, the problem of discovering the substrates of enzymes encoded by putative cytochrome P-450 genes may appear to be a little less daunting than the problem with protein kinases, but this is not the case. So far around 200 putative cytochrome P-450 genes have been identified in the genome of *Arabidopsis thaliana* (K. Feldman, personal communication). It may be worth noting in passing that, in spite of possessing so many cytochrome P-450 genes, *Arabidopsis* is sensitive to virtually every herbicide. Furthermore, in the only case where the endogenous and xenobiotic substrates of a specific cytochrome P-450 have been identified, the natural substrate is the simple, unsaturated linear compound lauric acid and the xenobiotic is the bicyclic aromatic herbicide diclofop. These two compounds are structurally extremely dissimilar, but this does not mean that the enzyme is so promiscuous that it will metabolise a wide range of herbicides (Helvig *et al.*, 1996). Substrate specificity in the cytochrome P-450 family is clearly an extremely complex phenomenon that will not readily yield to theoretical analysis on the basis of sequence data.

Protein Structures cannot be Predicted from Gene Sequences and do not Indicate Biochemical Functions

Another possible approach is to attempt to predict the three-dimensional folding patterns of proteins from the physicochemical properties of their polypeptide sequences and, since function is a property of structure, try to obtain some clues as to their possible functions in that way. Many of the proteins so far crystallised tend to fall into a limited number of groups, whose members share remarkably similar three-dimensional structures. There thus appear to be a limited number of stable conformations into which polypeptide chains can fold. Unfortunately, the members of a group which share extremely similar three-dimensional structures may have no detectable similarity in their gene or amino acid sequences (Doolittle, 1994). Additionally, many nascent proteins require the presence of molecular chaperones as they emerge from the ribosomes and substantial investments of ATP to ensure that they fold into functional conformations (Hartl, *et al.*, 1994). Their *in vivo* tertiary structure is, therefore, not entirely implicit in the primary amino acid sequence and could not be predicted from it with an acceptable degree of probability using purely physicochemical criteria. In any case, as far as determining function is concerned, it would probably not be a very useful exercise. One of the most common structural domains found in proteins is the TIM-barrel fold (Bränden, 1991). Polypeptides containing this domain are structurally so similar that 150 α-carbon atoms of any two members of the group can usually be superimposed to within a root mean square value of 2.5 Å. However, in addition to the complete absence of any detectable amino acid sequence similarity, such proteins have extremely diverse functions ranging from triose phosphate isomerase to flavocytochrome b_2. Three-dimensional structure is thus not a useful indicator of function.

Protein Structures Change when Ligands Bind

When the function of a protein is already known, knowledge of the three-dimensional structure is of great practical value in attempting to understand how, in mechanistic terms, the function is carried out. However, it should be noted that solving the three-dimensional structure of a protein *per se* is only of limited, speculative help in the discovery of new lead structures for potential inhibitors. Protein structure is dynamic and changes, sometimes quite considerably, but above all, unpredictably when a ligand binds. Only when a lead substance has already been discovered, probably by screening, and co-crystallised with its target protein, can molecular modelling come fully into its own.

Evolutionary Divergence can Radically Change the Sequences of Proteins which Retain Identical Functions

There are other significant problems in trying to use sequence data to determine function. For example, lysozyme can apparently tolerate many amino acid substitutions and still maintain its function. This has allowed such a high degree of sequence divergence during evolution that it would be difficult to predict, on the basis of sequence data alone, that some of the more distantly-related members of the group carried out the same biochemical function. With the benefit of hindsight, the divergent sequences yield a perfectly sensible evolutionary tree that is entirely consistent with other phylogenetic data, but this is only possible because the biochemical function of the enzyme is already known. There is no way of knowing how many more examples of this type there may be in the sequences of genes whose functions have yet to be determined. Furthermore, some biochemical functions of enzymes have arisen on more than one occasion during evolution (Doolittle, 1994). Consequently, polypeptides having no evolutionary relationship and thus neither structural nor sequence similarity can carry out identical biochemical functions. Again, only because the biochemical functions of the proteins concerned were already known were sequence data of any help in understanding this phenomenon; they could not possibly have predicted it.

When extremely similar protein structures can be formed by entirely different amino acid sequences and can have equally different functions, and identical biochemical functions can be carried out by proteins with entirely different sequences and structures, predictions of the functions of proteins on the basis of anything but quite high levels of sequence homology must be treated with great caution and, even then, subjected to rigorous biochemical and biological confirmation before they can form the basis of an industrial research project.

CONCLUSIONS

Crop protection research has an enormous advantage over pharmacology. It is free to be Darwinian. It can exploit the power of "natural selection", directly on the plants it aims to protect, to discover compounds with potentially useful biological activity in systems with very high levels of biological complexity. Prior knowledge of their precise biochemical interactions with the system is completely unnecessary. Only in this way have compounds with truly novel modes of action been discovered. Once biological activity has been detected, all the powerful tools that molecular biology offers can be brought to bear to provide a molecular rationale. Attempting the reverse, beginning with DNA and trying to achieve a predetermined biological effect, is rather like trying to push an object with a piece of string. If evolution were both directed and Lamarkian and the fore-runners of the giraffe had induced the heritable mutations required to make their necks grow longer by straining for the juicy leaves at the tops of trees, the method would perhaps have a greater chance of success.

One of the most unfortunate phases in the biological literature is "the central dogma." It was no doubt justifiable at the time to use any rhetorical device to emphasise that genetic information flows exclusively from DNA to protein and not *vice versa,* but the expression has a regrettable ring of certainty more suited to the utterances of an old testament prophet than to a humble, doubting scientist. Since then, the pronouncements of molecular biologist have tended to be accepted as carved in stone and the molecular genetic paradigm has come to dominate perceptions to a degree that is not always constructive. The elegant simplicity of molecular genetics has encouraged a view of biological systems which can grossly underestimate their vast complexity and adaptability. *The Blind Watchmaker* is mistakenly assumed to be tinkering with a watch, but biology is not clockwork. Within a single organism a dynamic web of immense complexity arises from the relatively simple rules of molecular genetics and this is continuously modified by equally dynamic inputs from other systems. All components are connected to, and affected by each other, and a change in one point of the system may have small or catastrophic consequences in another. The difficulties of predicting those consequences increase exponentially with the distance between the points. In these terms, the DNA of the genome and the macroscopic biology observed in the field could hardly be further apart. Such a labyrinth of complexity can absorb the research efforts of those who fail to acknowledge its shifting intricacy, quite indefinitely, and never yield practical results.

Improving our understanding of the world around us is a justified end in itself, but knowledge does not inevitably bring new solutions to practical problems. It would scarcely be possible to know more about the insulin molecule or its gene, but there is still no cure for diabetes, only a palliative treatment which existed long before molecular biology. Fortunately, the detailed understanding of problems is not a precondition for solving them. If it were, the material progress of humanity would have been very slow indeed. Primary solutions to problems are always pragmatic and empirically devised, whilst, to paraphrase Kierkegaard, understanding them is always retrospective. The advent of molecular biology and the sequencing of complete genomes do nothing to change this.

REFERENCES

Akers, A., Ammermann, E., Buschmann, E., Götz, N., Himmele, W., Lorenz, G., Pommer, E.-H., Rentzea, C., Röhl, F., Siegel, H., Zipperer, B., Sauter, H. & Zipplies, M. (1991) Chemistry and biology of novel amine fungicides: Attempts to improve the fungicidal activity of fenpropimorph. *Pesticide Science,* **31,** 521-38.

Ashman, W.H., Barbuch, R.J., Ulbright, C.E., Jarret, H.W. & Bard, M. (1991) Cloning and disruption of the yeast C-8 isomerase gene. *Lipids,* **26,** 628-32.

Baloch, R.I., Mercer, E.I., Wiggins, T.E.R. & Baldwin, B.C. (1984) Inhibition of ergosterol biosynthesis in *Saccharomyces cerevisiae* and *Ustilago maydis* by tridemorph, fenpropimorph and fenpropidin. *Phytochemistry,* **23,** 2219-28.

Bard, M., Lees, N.D., Turi, T., Craft, D., Cofrin, L., Barbruch, R., Koegel, L. &

Loper, J. C. (1993) Sterol synthesis and viability of *erg11* (cytochrome P-450 lanosterol demethylase) mutations in *Saccharomyces cerevisiae* and *Candida albicans*. *Lipids*, **28**, 963-67.

Bränden, C.-I. (1991) The TIM barrel - the most frequently occurring folding motif in proteins. *Current Opinion in Structural Biology*, **1**, 978-83.

Doolittle, R.F., Convergent evolution: the need to be explicit. (1994) *Trends in Biochemical Sciences*, **19**, 15-18.

Ellis, S.W., Rose, M.E. & Grindle, M. (1991) Identification of a sterol mutant of *Neurospora crassa* deficient in $\Delta^{14,15}$-reductase activity. *Journal of General Microbiology*, **137**, 2627-30.

Hartl, F-U., Hlodan, R. and Langer, T., (1994) Molecular chaperones in protein folding: the art of avoiding sticky situations. *Trends in Biochemical Sciences*, **9**, 20-25.

Helvig, C., Tardif. F.J., Seyer, A., Powles, S.B., Mioskowski, C., Durst, F. & Salaün J.-P. (1996) Selective inhibition of a cytochrome P-450 enzyme in wheat that oxidizes both the natural substrate lauric acid and the synthetic herbicide diclofop. *Pesticide Biochemistry and Physiology*, **54**, 161-71.

Johnston, M. (1996) Genome sequencing: the complete code for a eukaryotic cell. *Current Biology*, **6**, 500-03.

Loper, J.C. (1992) Cytochrome P-450 Lanosterol 14a-demethylase (CYP51). Insights from molecular genetic analysis of the *ERG* 11 gene in *Saccharomyces cerevisiae*. *Journal of Steroid Biochemistry and Molecular Biology*, **43**, 1107-16.

Ladevèze, V. Marcireau, C. Delourme, D. & Karst, F. (1993) General resistance to sterol biosynthesis inhibitors in *Saccharomyces cerevisiae*. *Lipids*, **48**, 907-12.

Marcireau, C., Guyonnet, D. & Karst, K. (1992) Construction and growth of a yeast strain defective in sterol 14-reductase. *Current Genetics*, **22**, 267-72.

BIOPHARMACEUTICALS IN THE LATE GENOMIC ERA

RICHARD A.G. SMITH

AdProTech plc, Unit 3, Number 2 Orchard Road, Royston, Herts SG8 5HD, UK

Why the *late* genomic era? Why not the *post*-genomic one so frequently proclaimed? A first response is to question whether, from the standpoint of obtaining new insights into biology generally, we really have now grasped all that large-scale sequencing can offer. After all, the (majority) non-expressed parts of human and other genomes are still far from fully sequenced, the vast question of gene control in development is only now being addressed and the sequence/structure and structure/function questions remain (albeit illuminated at the margins by much larger databases). A more positive point, relevant to a therapeutic theme, is that while functional genomics may have a long way to go, in the post-sequencing era we are at least able to assess the intra- and inter-genomic *diversity* of genes and their protein products. The role of that diversity in the manifestations of disease and its relationship to the concept of chemical diversity (now so critical to drug discovery) form the background to this paper.

The term 'biopharmaceutical' is a persistently woolly one. It is sometimes applied to all therapeutics derived by biological processes but this would include beta-lactam antibiotics which lie firmly in the chemical pharmacopoeia. A more 'chemical' definition speaks of drugs with large and complex molecular structures but just how 'large' and how 'complex'? Here, biopharmaceuticals will be defined as therapeutic agents derived from natural biological macromolecules by extractive, recombinant and chemical techniques. This definition is still woolly (how 'macro' is 'macro'?) but it does have the virtue of being comprehensive and encompasses nucleic acids, including antisense, gene therapeutics, chemically modified DNA and perhaps peptide nucleic acid and aptamers, protein therapeutics (including engineered and 'minimised' structures) and complex carbohydrates. It is immediately obvious that the output of the late genomic (= post-sequencing) era directly impacts only some of these categories. Engineered structures depend on structure-function input and comparative pharmacology of primary gene products. Gene therapy and antisense are critically dependent on functional genomic insights. Carbohydrates lie in the post-translational, biosynthetic world. Therefore, I shall focus mainly on primary protein/gene therapeutics and those of their derivatives that can be immediately derived from genomic relationships.

In this context, the potential diversity of biopharmaceuticals can be viewed at three levels. At the primary genomic level, there is the specific question of how many expressed human genes possess therapeutic potential in man . To this can be added the often overlooked enumeration of therapeutically useful genes and gene products from other species and the even less explored area of possibly useful products from the non-expressed genome (e.g. pseudogenes). The second level of diversity is the structural one. A wide range of autonomously folded protein structural domain types exist (for a discussion of just how many, see Jones *et al.,* 1992). Combinatorial assembly of domains to create products with multiple or targeted functions is well documented. For example, the blood coagulation and fibrinolytic systems contain mosaic proteins in which a protease function may be modulated by non-catalytic domains (such as growth factor domains or kringles) which permit interactions with cell-surface receptors or structural proteins and thus localise and control enzymatic activity. This combinatorial generation of novel function is the product of evolutionary selection of useful products from exon-shuffling events, but the creation of novel combinations by directed intra- and inter-genome domain selection offers wide (and as yet, minimally exploited) possibilities for the design of therapeutics. The final level of diversity is that of sequence variation within folded domains to create novel binding specificities. The potential in this area is well illustrated by just one domain – the F_{ab} (scF_v) of immunoglobulins. Here, variation of sequence within the complementarity determining regions (CDRs) is the origin of most antibody diversity and hence the specificity of the immune response and has also given rise to the burgeoning speciality of antibody engineering (Winter & Milstein, 1991; Reff, 1993; Jung *et al.*, 1997)

A perspective on the potential diversity of the biopharmaceutical pharmacopoeia can be provided by some simple calculations. Let us assume that the recently developed techniques in 'signal trapping' (Tashiro *et al.,* 1993; Klein *et al.,* 1996) together with conventional large-scale sequencing have enabled us to identify a set of around 6000 genes encoding secreted proteins. At the current state of understanding of function, let us also assume that 10% of these proteins have sufficient biochemical relevance to disease to be regarded as *potentially* of therapeutic value (possible transfer of partial function or transformation of agonist to antagonist pharmacology or *vice versa* being considered to be admissible). Then we can add perhaps 100 useful gene products derived from non-secreted and other sources to give a portfolio of around 700 proteins and 1000 domains. The conservative nature of the latter figure will be apparent from the discussion on short consensus repeats below. The number of possible hypothetical gene/protein therapeutics consisting of <u>two</u> domains chosen for their particular combinations of binding specificities, allosteric modulators, enzymatic activities etc is therefore around 10^6. Permit your domain-shufflers to spread their wings to create 3- or 4-domain constructs and resultant diversity increases significantly. Now we can factor in sequence variation to 'fine-tune' binding specificities. Assume that the average 'domain' is around 80 amino acids, that around 80% of these are required to maintain the overall domain architecture and that of the remainder about half can be manipulated to change specificity: the potential diversity of the mutated domains comes out in the region of 10^{10} – a figure that will seem familiar from phage-display experiments with individual domains (Smith & Scott, 1993). Thus, the *total* diversity of therapeutic biomolecules (whether used as genes or proteins) which are derivable by rational design/targeted mutation steps is <u>at least 10^{16}</u> – a figure that should give combinatorial chemists struggling to validate 10^6-mer libraries pause for thought. Of course, there are constraints on the use of biopharmaceuticals which are discussed in more detail below but, even if we assumed that only 1 in 1000 possible proteins could be designed to meet the required therapeutic profile, the scope for creating novel proteins that are plausible drug candidates is still substantial.

For credibility, these sums require a specific illustration. In this laboratory we have a strong interest in anti-inflammatory therapeutics based on modulators of the complement system. The latter cannot be reviewed in detail here – several excellent texts are available (Morgan, 1990; Dodds & Sim, 1997). Suffice it to say that complement provides an effector mechanism for converting the binding specificity of an antibody/antigen interaction into a transport/removal and cytolytic response (Classical Pathway) and also constitutes a first line of defence against invading organisms by providing a molecular self/non-self recognition mechanism (Lectin and Alternative Pathways). Contrary to popular belief, complement is not just a backwater of immunology beset by arcane nomenclature but is central to the initiation and progression of wide variety of pathogenic inflammatory processes in conditions as diverse as myocardial infarction and rheumatoid arthritis. Complement activation triggers chemotactic, opsonisation, haemostatic and nuclear activation phenomena which integrate the activation pathways with many other pathological processes. The activation of complement is largely controlled by the products of the Regulators of Complement Activation (RCA) gene cluster and the therapeutic use of the RCA proteins has been reviewed (Mossakowska & Smith, 1997; Makrides, 1998).

Prominent among these regulatory molecules is human complement receptor 1 (CR1,CD35, C3b/C4b receptor, Fearon, 1979; Sim, 1985). In its most common allotypic form, CR1 consists of no less than 30 linear repeats of a 60-70 amino acid 2-disulphide domain known as the short consensus repeat (SCR, CCP or Sushi domain) followed by a single transmembrane domain and a short cytoplasmic region. CR1 binds and transports immune complexes (antigens complexed to specific immunoglobulins and containing covalently bound complement proteins, primarily C3b). It also acts as a template upon which the protease intermediates of the complement system (the convertases) are broken down and forms a link (through interactions with another complement receptor, CR2, and other B-cell membrane proteins) between complement activation and the control of immunoglobulin synthesis.

This extravagantly elongated multifunctional receptor is, from several points of view, a protein engineer's dream. Firstly, the SCR domains can be organised on the basis of homology into 4 regions of 7 SCRs each (long homologous repeats), the first three of which contain functional binding sites for the activation intermediates C3b and C4b. Secondly, the SCR domain family contains at least 180 different members (1998 public databases) and it is likely than the human genome has at least 100 unique SCR sequences and perhaps as many as 300. Thirdly, the structures of individual SCR domains and pairs of domains from a related protein (Factor H) have been determined by NMR methods (Barlow et al., 1991, 1993). From this it is known which intra-domain loops within the ~70% of the SCR sequence that is not conserved between family members are likely on structural grounds to be involved in protein-protein interactions. Thus, from the standpoints of known functionality, domain-shuffling potential and intra-domain sequence variation, CR1 is a paradigm for the biopharmaceutical diversity of the wider genome. We have explored a tiny fraction of the possible variants by expression of functional CR1 fragments in bacteria (Dodd et al., 1995) and have extended these studies to include mutagenesis and inter-domain sequence exchanges. Such studies suggest that the complement inhibitor function of CR1 can be incorporated in structures <10% of the size of the parent molecule and that directed mutagenesis and targeting can greatly increase the potency and selectivity of this activity. When one realises that the wild-type extracellular region of CR1 (soluble CR1, sCR1, TP10) has now been shown to be active in preventing reperfusion damage and transplantation rejection (Weisman et al., 1990; Ryan, 1995), the implications of the vast scope for therapeutically focussed improvement of this type of biopharmaceutical become clearer. What holds for an SCR-based therapeutic also holds for other structurally-based families. For example, the 4-helix bundle cytokines include such mediators as erythropoeitin, interleukins 4,5 and 6 and leptin, engineered analogues of which may address therapeutic targets as diverse as anaemia and obesity. The CC and CXC cytokines are further examples of large and growing families of soluble secreted proteins whose biological effects depend on a particular spectrum of receptor interactions which may be susceptible to rational manipulation through redesign of the domain archetype.

It is now necessary to consider what practical constraints exist on the use of biopharmaceuticals and what prospects there are of overcoming them. Broadly speaking, proteins and genes owe their (relative) unpopularity among drug developers to three main factors:

1. Bioavailability and biodistribution: proteins and nucleic acids are generally too large to be absorbed directly and do not survive digestive processes. Consequently they have to be given by parenteral routes (usually intravenously). Except for cases where self-administration is possible (e.g. insulin in diabetes) or dosing is relatively infrequent (e.g. interferon-β in multiple sclerosis), this restricts the duration of therapy so that important chronic indications are not accessible. The size of proteins and nucleic acids also often denies them access to extravascular compartments, to tissues and to intracellular targets including the nucleus.

2. Potency and cost of goods: relative to most synthetic chemicals, biological molecules are still expensive to manufacture despite broad advances in process biotechnology. The high costs derive not only from the capital investment needed for fermentation and isolation plant but also from the difficulties of characterising and controlling complex molecules for therapeutic use. As a rough guide, economically desirable gravimetric doses in man are at least 10- fold lower than for chemical entities. This in turn creates a need for molar potencies up to 10^3-fold higher than for low-MW drugs. Where nanomolar potency may be the target in large-scale screening of chemical libraries, efficacy at picomolar blood levels may be needed for a biopharmaceutical.

3. Immunogenicity and immunotoxicology: although small and very highly purified proteins can show minimal immunogenicity in man even upon repeat administration, the reality is that most protein preparations (even those with fully human sequences and apparently 'native' structure) will elicit a specific antibody response in humans. In the majority of cases, this response is non-neutralising either because the epitopes involved are distant from the key regions of the protein or because the actual immunogen is a minor

impurity or misfolded product which is irrelevant to the action of the drug substance. However, the possibility that there will be an immediate or longer-term harmful immune response always needs to be considered with macromolecular therapeutics, especially proteins and especially those with intentionally altered sequences.

The availability of new sequence data has several interrelated effects on these barriers. Firstly, if the biodistribution (disease-site targeting) of an agent can be improved by conferring another function on it through fusion to another binding element, so usually can its potency and economics of production. If the effect is large enough, it may be possible consider alternative routes of administration – an agent active at picomolar levels in blood may only need to be absorbed with 0.1% efficiency (for example by the intranasal route) to be effective at economic doses. Such efficient targeting is possible: a recent study showed that linking the human C3d protein sequence in multiple copies to a model antigen enabled an immune response at doses in mice which correspond to sub-microgram amounts in man (Dempsey *et al.*, 1996). This was achieved because the antigen was targeted to B-lymphocytes involved in the synthesis of specific immunoglobulin. Potency may also be increased by changes which improve stability and resistance to proteolytic degradation *in vivo*, by post-translational modification and by sequence optimisation based on structural analysis of related members of a protein family. Improvements in databases also impact the immunogenicity issue by defining more clearly what sequences are human and what the structural context of engineered non-human sequences is likely to be. So, for example, if a novel hybrid protein contains a junction sequence created from two human components this can be optimally 'humanised' by reference to the expressed human genome in general and to the family of proteins in particular. It is also possible to use sequences known or predicted to be flexible or derived from non-human proteins of very low immunogenicity (such as occur, for example, in parasites). Such changes may also be made in regions distant from the active sites of the components so that any epitope-specific immune response is non-neutralising.

If the generic disadvantages of biopharmaceuticals are being addressed by improved protein design as well as gene discovery, it is worth restating that the main generic advantages of these molecules include high specificity for a molecular target and a usually benign toxicological profile due to the intrinsically low toxicity of polypeptides and polynucleotides. The general toxicity of protein therapeutics is usually predictable from their pharmacological properties. The ability of biomolecules to target specific mediators in disease may be critical to effective and safe treatment, but there is a danger that focussing on a single mediator or process may lead to therapeutic failure where multiple pathological processes are involved. An important example of this failure of therapeutic reductionism seems to have occurred in the treatment of septic shock.

The shock states cover a range of syndromes including adult respiratory distress syndrome (ARDS) , systemic inflammatory response syndrome (SIRS), disseminated intravascular coagulation (DIC) and multiple organ failure (MOF) – the latter being a common end point in these processes. Shock results from the uncontrolled release of a wide variety of mediators from activated cells which give rise to myocardial and pulmonary dysfunction, cardiovascular collapse, hypotension, organ failure and death. The initiating factors are often derived from infectious agents (e.g. lipopolysaccharide from Gram-negative bacteria) but shock states can also follow chemical or thermal burns or other trauma where there is no overt bacteremia. The response to the initial stimulus may involve activation of platelets, neutrophils and macrophages with release of a range of cytokine and chemokine mediators including interleukins 1, 6 and 8, tumour necrosis factor (TNF), platelet activating factor (PAF) and products of coagulation and complement activation. Antibodies to or antagonists of many of these mediators or their receptors have been identified but the results of clinical trials with such agents have been remarkably disappointing with little reduction (or in some trials an increase) in mortality. Apart from the failure to address an area of major unmet medical need (US cases of septic shock exceed 300,000 per year with an overall mortality of ~25%), the trials of single mediator antagonists have cost hundreds of millions of dollars and their failure has cast a shadow over the entire biotechnology industry.

The experience of septic shock suggests that targeting single mediators is misguided because by the time the various shock syndromes are diagnosable, the inflammatory processes are being maintained by multiple pathways involving positive feedback loops. Although several ways of tackling this situation can be envisaged (for example at the gene transcription level) it does raise the specific question of how frequently multifunctional biopharmaceuticals may be needed to address complex pathological processes with redundant and/or interacting pathways. Nature has clearly devised multifunctional proteins for biosynthetic and regulatory purposes and the kind of domain-shuffling referred to above makes such molecules quite feasible. However, the construction and efficient production of (for example) a heterospecific IgM-like molecule or a hybrid of a receptor antagonist, a protease inhibitor and an Fab domain is far from trivial. Innovative approaches such as using a generic low-toxicity linear template to assemble the desired combination of binding specificities post-translationally, may be needed to access such agents cost-effectively.

In summary, the end of the sequencing era marks the true end of the 'hunter-gatherer' phase of biopharmaceuticals. Here and there, good examples of domesticated, designed therapeutic macromolecules have appeared over the past decade and are supporting healthy patients and healthy balance sheets. A completely rescaled opportunity now exists to apply molecular evolution and screening techniques as well as 'mix and match' strategies to the design of architecturally complex but conceptually simple and specific protein molecules and their genes. Functional genomics will undoubtedly yield unique drug targets for which ligands will be designed and yield useful drugs. It would be surprising, however, if the 'one gene – one disease ' paradigm was to prove universal or even dominant. A rational approach to the complexities of multi-gene pathology will be needed and the future for better biopharmaceuticals looks bright.

REFERENCES

Barlow, P.N., Baron, M., Norman, D.G., Day, A.J., Willis, A.C., Sim, R.B. and Campbell I.D. (1991) Secondary structure of a complement control protein module by two-dimensional ^{1}H NMR. *Biochemistry* **30**, 997-1004.

Barlow, P.N., Steinkasserer, A., Norman, D.G., Kieffer, B., Wiles, A.P., Sim, R.B. and Campbell I.D. (1993) Solution structure of a pair of complement modules by nuclear magnetic resonance. *Journal of Molecular Biology* **232**, 268-284.

Dempsey, P.W., Allison, M.E.D., Akkarju, S., Goodnow, C.C. and Fearon, D.T. (1996) C3d of complement as a molecular adjuvant: Bridging innate and acquired immunity. *Science* **271**, 348-350.

Dodd, I., Mossakowska, D.E., Camilleri, P., Haran, M., Hensley, P., Lawlor, E.J., McBay, D.L. Pindar, W. and Smith R.A.G. (1995) Overexpression in Escherichia coli, folding, purification and characterization of the first three short consensus repeat modules of human complement receptor type 1. *Protein Expression and Purification*. **6**, 727-736.

Dodds, A.W. and Sim R.B. (1997) *Complement: a Practical Approach*. Series Eds. Rickwood, D. and Hames, B.D. Oxford, IRL Press.

Fearon D.T. (1979) Regulation of the amplification C3 convertase of human complement by an inhibitory protein isolated from human erythrocyte membrane. *Proceedings of the National Academy of Sciences of the USA* **76**, 5867-5871.

Jones, D.T., Taylor, W.R. and Thornton J.M. (1992). A new approach to protein fold recognition. *Nature.* **358**, 86-89.

Jung, S. and Pluckthun A. (1997) Improving *in vivo* folding and stability of a single-chain Fv antibody fragment by loop grafting. *Protein Engineering,* **10**, 959-966.

Klein, R.D., Gu, Q., Goddard, A. and Rosenthal. A. (1996) Selection for genes encoding secreted proteins and receptors. *Proceedings of the National Academy of Sciences of the USA.* **93**, 7108-7113.

Makrides S.C. (1998) Therapeutic inhibition of the complement system. *Pharmacological Reviews.* In Press.

Morgan B.P. (1990) *Complement: Clinical Aspects and Relevance to Disease*. London, Academic Press.

Mossakowska D.E. and Smith R.A.G. (1997) Complement receptors and their therapeutic applications. *Recombinant Cell Surface Receptors: Focal Point for Therapeutic Intervention.* Ed. Browne, M.J., Georgetown, Texas, R.G. Landes & Co. pp 209-220.

121

Reff, M. (1993) High-level production of recombinant immunoglobulins in mammalian cells. *Current Opinion in Biotechnology.* **4,** 573-576.

Ryan, U.S. (1995) Complement inhibitory therapeutics and xenotransplantation. *Nature Medicine* **1,** 967-968.

Sim, R.B. (1985) Large-scale isolation of complement receptor type 1 (CR1) from human erythrocytes. *Biochemical Journal.* **232,** 883-889.

Smith, G.P. and Scott J.K. (1993) Libraries of peptides and proteins displayed on filamentous phage. *Methods in Enzymology,* **217,** 228-257.

Tashiro, K., Tada, H., Heilker, R., Shirozu, M., Nakano, T. and Honjo. T. (1993). Signal sequence trap: a cloning strategy for secreted proteins and Type 1 membrane proteins. *Science.* **261,** 600-603.

Weisman, H.F., Bartow, T., Leppo, M.K., Boyle, M.P., Marsh jr, H.C., Carson, G.R., Roux, K.H., Weisfeldt, M.L. and Fearon D.T. (1990) Recombinant soluble CR1 suppressed complement activation, inflammation and necrosis associated with reperfusion of ischemic myocardium. *Science.* **249,** 146-151.

Winter, G. and Milstein C. (1991) . Man-made antibodies. *Nature.* **349,** 293-299.

COMMERCIAL STRATEIGIES FOR EXPLOITING GENOMICS

IDENTIFICATION OF NOVEL TARGETS IN GENE SEQUENCE DATABASES

M. J. BROWNE

Biopharmaceutical Research UK, SmithKline Beecham, New Frontiers Science Park, Third Avenue Harlow, Essex, CM19 5AW, UK

INTRODUCTION

Recombinant proteins provided a new source of therapeutic targets and agents during the 1980s and early 1990s. However, during the early 1990s, it became apparent that the identification and cloning of novel cDNAs was a rate-limiting step in drug discovery and that new technological approaches were required to address the challenge. In 1993, we formed an alliance with Human Genome Sciences to create a large database of expressed sequence tags (ESTs, Adams *et al.*, 1995), this has greatly expanded our ability to recognise novel therapeutic targets. The productivity of this strategy is illustrated by reference to our work on novel enzymes, chemokines and receptors, and new approaches linking genes to pathological processes.

DATABASE TO TARGET

The effective exploitation of the dataset depends on the rationale for cDNA library construction, the biological/structural reasoning driving the searches and the sophistication of the bioinformatic tools employed. The SB/HGS database originates from several hundred cDNA libraries derived from many different, normal and diseased, human tissues and a wide variety of cell types. In the past four years, we have invested in a major initiative in bioinformatics to organise and interrogate this information (Marshall, 1996). We have attempted to integrate the EST database into all aspects of drug discovery. In the following sections examples are provided to illustrate how the database has been of central importance to our drug discovery programmes.

At the outset of this project, it was clear that creation of an EST database was only the start of the process to design new effective and selective therapies. The next key issue was identifying within the database those genes that were most relevant to the drug discovery process.

Atherosclerosis.
Atherosclerosis is a major cause of morbidity and mortality. Despite recent therapeutic progress, eg in the area of HMGCoA reductase inhibitors, it is clear that further strategies to control pathological events, especially in vascular wall biology

are required. We were, therefore, interested in identifying other potential targets. Two approaches are described; both fundamentally rely on hypothesis-driven interrogation of the sequence database, and either employ data from protein microsequencing projects, or a search for homology to known genes.

Novel lipoprotein-associated phospholipase. Oxidation of low-density lipoprotein (LDL) and production of lysophosphatidyl choline is a key event in the development of atherosclerosis. Lysophosphatidyl choline is a powerful chemoattractant for monocytes and also induces expression of endothelial leukocyte adhesion molecules - inhibition of its formation might, therefore, be expected to have therapeutic benefit. Hence we began a search for the phospholipase responsible for generating lysophosphatidyl choline. It was known that plasma PLA_2 levels are raised in familial hypercholesterolaemia and this provided the starting point for what initially began as a conventional cDNA cloning project. A PLA_2 activity was purified from patient plasma, and microsequencing of this protein enabled a classical oligonucleotide cDNA library probing strategy to be initiated. However, the project was considerably accelerated 'midstream' as the EST database became available and identification of the cognate cDNA was accomplished in only a few minutes (Tew *et al.*, 1996). The enzyme was then overexpressed using a baculovirus system and used as the basis for a high throughput screen for potent inhibitors.

Though the gene was identified via a microsequencing project, we noted that the chromosomal breakpoint in a family with autosomal dominant supravalvular aortic stenosis falls within the 5' untranslated region of the LDL-PLA_2 gene, lending further, indirect support for an association between this novel protein and cardiovascular pathology.

A novel CC chemokine. Chemokines are a family of small proteins of 8 - 10 kDa in size which fall into at least 2 classes, defined by the arrangement of disulphide bonds, the CxC (or alpha) and the CC (or beta) chemokines. Chemokines have been implicated in a number of inflammatory processes associated with pathology. Of particular interest to atherosclerosis, are the CC chemokines, since the class prototype MCP-1 is chemotactic for monocytes, and infiltration of monocytes is implicated in the initiation and development of atherosclerosis and stroke. Homology-based searching of a foetal brain cDNA library yielded a novel member of the CC chemokine family, now termed MCP-4 (Berkhout *et al.*, 1996). Expression, using a *Drosophila* cell system, produced a protein capable of selective, potent monocyte chemotaxis *in vitro*, highlighting a potential role for this novel chemokine in inflammatory disease and its potential for drug discovery.

Osteoporosis.
Tissue-specific expression of genes can provide clues to their role in pathology. One method of searching for differences in gene expression is to interrogate a large

collection of ESTs from a variety of cDNA libraries ('electronic' Northern).

<u>Osteoclast-specific cysteine proteinase (Cathepsin K).</u> Osteoporosis is an increasingly important disease in an ageing population. Normally bone is continually remodelled, however, when this process fails to balance correctly between bone resorption and regeneration, bone loses mass and becomes brittle. We were interested in finding means to prevent this breakdown of bone.

Osteoclasts are implicated in bone resorption. A cDNA library was constructed from an osteoclastoma library and was EST-profiled. Bioinformatic searches revealed an abundant transcript that was present as 4% of all ESTs and, by homology, identified as a novel member of the cysteine proteinase class (Drake *et al.*, 1994). This novel gene was shown by *in situ* hybridisation to be expressed in osteoclasts but not osteoblasts, confirming database information on specificity of tissue distribution. Subsequent immunohistochemical studies also confirmed the localisation of expression. We are currently evaluating the therapeutic potential of this target in bone resorption by making specific small molecule inhibitors.

Whilst we had identified the cathepsin K gene by the 'database-first' strategy, it is interesting to note that the critical role of this gene in bone biology has been supported by the work of Gelb *et al.*, (1996) in a genetic-based study of Pycnodysostosis. This is a rare autosomal recessive disease manifested by osteosclerosis and short stature. Gelb *et al.*, were able to locate defects in the *same* gene in 3 ethnically disparate groups.

<u>Seven transmembrane G-protein coupled receptors</u>

The seven-transmembrane receptors (7tms) hold a unique place in the armamentarium of the pharmaceutical industry - even members of the general public are familiar with 'beta-blockers'. Many new members of the 7tm family were discovered, in the last decade, using low stringency probing and/or PCR methodologies. Although highly successful, this approach is limited by the signal to noise ratio inherent in hybridisation or PCR-based strategies, by contrast, homology searches *in silico* can access a wider range of sequence divergence.

We have identified many dozens of new receptors via the EST paradigm. Once a potential receptor has been located, it is necessary to define function. Using *Xenopus* oocyte and mammalian cell expression systems we have already been able to identify receptors for calcitonin gene related peptide (CGRP) and the complement C3a component (an important mediator of the inflammatory response).

Despite these successes, the number of new receptors poses a considerable scientific and logistical challenge, and many cannot currently be utilized in drug discovery since their functions, and thus their potential therapeutic relevance, are unknown. We have therefore added a novel yeast-based system to our overall strategy via a collaboration with Cadus. The Cadus technology is designed to enable rapid expression and *functional* coupling of human cellular receptors in yeast. This technology will simultaneously provide tools for discovering receptor function and immediately create high-throughput screens to discover chemical compounds -- potential drugs -- that interact with these receptors. The SB/Cadus collaboration thus creates a bridge between genomics and high-throughput screening for discovery of new agonists or antagonists, and has the potential to rapidly mine valuable new discovery opportunities from a previously underexploited source.

LINKING THE EST DATABASE TO BIOLOGY

The examples given above have focused on genes that have been identified through 'electronic probing' of the database for a match to a novel peptide sequence, or for homology with known protein families, or searching for genes which have a selective tissue distribution. An alternative approach is to use this large gene collection directly to interface with genomic DNA (positional cloning) or mRNA (expression profiling) projects, in other words, 'biological probing' to identify disease genes for use either as therapeutic targets or as diagnostic markers.

Positional and Positional Candidate Cloning Strategies

A critical stage in any positional cloning programme is the construction of a contig of YACs/PACs. Typically, identification of transcripts arising from this contig is highly resource intensive, yet there is a high probability that any 'disease gene' may already be represented in the EST database. The key problem is how to interface these two resources. We are, therefore, investigating solid phase presentation methods for arraying large numbers of clones in a form accessible to segments of genomic DNA identified in positional cloning projects. Early results with PAC's and YAC's from both the early onset Alzheimer's Disease locus on chromosome 14 and the Pycnodysostosis locus are highly encouraging and bode well for the prospects of creation of a non-redundant human cDNA array as a convenient, potent, reusable resource for transcript identification.

As an alternative 'disease gene' cloning route Francis Collins (1995) has advocated the 'positional candidate' strategy. This strategy becomes more powerful as the density of transcripts and other markers on the human genome increases.

In order to exploit the positional candidate strategy, it is necessary to assign genes of interest to candidate loci. Traditionally, mapping of candidate genes has often been expensive and inaccurate. With the recent advances in gene mapping techniques mapping of cDNAs and ESTs has become much more accessible (Hudson *et al.*,

1995; Gyapay *et al.*, 1995). We have used both fluorescent *in situ* hybridisation (FISH) and Genebridge 4 radiation hybrid mapping techniques. Numerous genes have been mapped, pointing up potentially significant association with important disease loci.

The value of the database to locate 'disease genes' is illustrated by our work on human galactokinase.

Cataracts are the leading cause of blindness in the world. The aetiology of most cataracts is unknown, but homozygous deficiency of galactokinase is a known cause of cataracts. Purification and microsequencing of the human protein enabled rapid cloning of a novel cDNA following a database search (Stamboulian et al., 1995). The cDNA was shown to encode a functional galactokinase by a variety of methods including complementation of galactokinase-deficient *E coli*. Having shown biochemical activity it was then possible to link this specific gene to the human genetic defect: the cDNA was mapped by fluorescent *in situ* hybridisation to chromosome 17q24 (the known susceptibility locus).

We then looked at the galactokinase gene carried in two families where there was a known galactokinase deficiency. In family AC we were able to show very clearly there was an in frame stop codon at aa80 and in family GB there was amino acid substitution at position 32 with a transposition from valine to methionine. The latter protein, when expressed *in vitro*, was shown to be completely lacking in enzymatic activity. We were able to investigate both heterozygous and homozygous members of the families and only in the homozygous state was the gene associated with a obvious metabolic defect.

In line with our early results we anticipate this will become an increasingly fruitful strategy.

Gene Expression Monitoring

High-density gene micro-arrays also have considerable potential for monitoring of gene expression patterns in normal and diseased tissues. Using our large set of cloned human ESTs we are currently evaluating high-throughput instrumentation designed to increase the sensitivity of the technology and speed the generation of data. Of particular importance is the ability to generate a sufficiently large data set to be able to distinguish between inter-individual variation and the variation underlying overt pathology.

MICROBIAL GENOME SEQUENCING

In addition to our work on human genes, we are also applying similar high throughput sequencing strategies to the search for novel antimicrobial targets. This is becoming an increasingly important area for research: The World Health Report 1996 reveals that infectious diseases kill over 17 million people a year. Drug-resistant strains of microbes are increasing and there is a critical need for new products to treat the emerging threat of these micro-organisms. Putting this in perspective, the last new antibiotic class was discovered in the mid-1970's. Unravelling the genetic sequence of the microbial agents that cause disease offers a potent new route for developing new therapies.

FUTURE PROSPECTS

The EST-centred cloning and bioinformatic strategy has yielded tangible impacts for the drug discovery process already. It can be anticipated that this trend will continue both for cloning and diagnostic applications. This will continue to be complemented with technologies for directly interfacing novel cloned genes with biological systems, to determine which enzymes or receptors are important in a particular disease process and which may be beneficially activated or inhibited by pharmacological intervention. Moreover, improved linking between therapeutics and diagnostics will permit both more selective targeting of therapy in the clinic and more rapid establishment of 'proof of concept' in clinical trials.

REFERENCES

Adams, M.D., Kerlavage, A.R., Fleischmann, R.D., Fuldner, R.A., Bult, C.J., Lee, N.H., Kirkness, E.F., Weinstock, K.G., Gocayne, J.D., White, O., (1995) *Nature, Genome Directory*, **377,** supp, 3-174.

Berkhout, T.A., Sarau, H.M., Moores, K., White, J.R., Elshourbagy, N., Appelbaum, E., Reape, T.J., Brawner, M., Makwana, J., Foley, J.F., Schmidt, D.B., Imburgia, C., McNulty, D., Matthews, J., O'Donnell, K., O'Shannessy, D., Scott, M., Groot, P., Macphee, C., (1997) Journal of Biological Chemistry, **272,** 16404-16413.

Collins F S (1995) *Nature Genetics*, **9**, 347-50.

Drake, F., James, I., Connor, J., Rieman, D., Dodds, R., McCabe, F., Bertolini, D., Barthlow, R., Hastings, G., Gowen, M., (1994) *J Bone Min Res*, **9**, S177.

Gelb, B.D., Shi, G.P., Chapman, H.A., Desnick, R.J., (1996) *Science,* August 30, **273**, 1236-1238.

Gyapay, G., Schmitt, K., Fizames, C., Jones, H., Vega-Czarny, N., Spillett, D., Muselet, D., Prud'Homme, J.F., Dib, C., Auffray, C., Morissette, J., Weissenbach, J., Goodfellow, P.N., (1995) *Human Molecular Genetics,* **5**, 339-346.

Hudson, T., Stein, L., Gerety, S., Ma, J., Castle, A.B., Silva, J., Slonim, D.K., Baptista, R., Kruglyak, L., Xu, S.H., (1995) *Science,* **270**, 1945-1954.

Marshall, E., (1996) *Science,* **272**, 1730-1732.

Stamboulian, D., Ai, Y., Sidjanin, D., Nesburn, K., Sathe, G., Rosenberg, M., Bergsma, D.J., (1995) *Nature Genetics,* **10**, 307-312.

Tew, D., Southan, C., Rice, S. Q., Lawrence, M.P., Li, H., Boyd, H.F., Moores, K., Gloger, I.S., Macphee, C.H., (1996) *Atherosclerosis, Thrombosis and Vascular Biology,* **16**, 592-599.

MICROBIAL GENOMICS AND THE DISCOVERY OF NEW ANTIMICROBIAL THERAPIES

M. EGERTON, A. E. ALLSOP AND G. J. BOULNOIS

ZENECA Pharmaceuticals, Alderley Park, Cheshire, SK10 4TG, UK

ABSTRACT

The sequencing of entire microbial genomes has revolutionised anti-microbial research. Anti-microbial drug discovery has traditionally relied upon random screening or semi-rational modification of known structural series. These strategies have failed to deliver the new agents required to address the infectious disease challenges currently being experienced in the clinic. Microbial genomics and associated genetic tools have provided a platform for the rational identification of new molecular targets. These technologies are transforming antibacterial and antifungal drug hunting, which should result in a wide variety of new antimicrobial agents with novel mechanisms of action.

INTRODUCTION

The use of antimicrobial drugs to control infectious diseases represents one of the greatest achievements of medicine this century. The versatility shown by microorganisms in overcoming the effects of these drugs, however, is no less remarkable. Bacteria and fungi have developed a variety of resistance mechanisms to every class of antimicrobial agent used in the clinic. Moreover, in many cases of resistant bacteria, resistance determinants have been transferred between strains and species (Gold and Moellering, 1996; Levy 1992; Neu, 1992; Davies, 1994). The pharmaceutical industry has been relatively successful in developing new drugs to address the problem of drug-resistant pathogens. However, increasing numbers of multiple-resistant isolates are now beginning to appear, that are causing serious problems in both the hospital environment and the community (WHO 1996). As a result the effective use of many major classes of drug has been severely limited (Rockerfeller University Workshop, 1994, Report of the ASM Task force on Antibiotic Resistance, 1995).

Development of resistance to antimicrobial agents will always be a concern, however, new drugs with novel mechanisms of action can potentially provide effective therapy for a period of time (Chopra *et al.*, 1996). Since the spread of resistance is largely dependant upon selection in the clinic, that is, use of the drug, a wider variety of novel agents would expand the armoury, increase the choice, and so prolong effective clinical utility of specific drug classes. In recent years many pharmaceutical companies have re-focused drug discovery away from new analogues of known drug classes and towards developing novel agents with new mechanisms of action. Numerous potential drug targets have been investigated but the scope of these investigations has been limited by our understanding of microbial molecular and cellular biology. The advent of microbial genome sequences has provided the information to begin to explore areas of science beyond our current understanding. The objective of this paper is to illustrate how microbial genomics can be exploited in the discovery of new antimicrobial agents and to identify some of the important opportunities that may be available in the future.

MICROBIAL GENOMICS AND DRUG DISCOVERY

The process of drug discovery can be broadly defined into three overlapping phases of Target Identification and Target Selection (TI/TS), Lead Identification (LI) and Lead Optimisation (LO) (Fig. 1). In the TI/TS phase, activities are focused on the identification of a molecular target that is suitable for development into a drug discovery programme, and the development of appropriate screening assays for use in later phases. The LI phase is focused on the identification of molecules that modulate the activity of the target in an appropriate way (e.g. enzyme inhibitors, receptor agonists or antagonists). This process often involves the screening of proprietary compound collections that usually contain many thousands of samples. The compounds identified in this process will not possess all the attributes that are required of the final drug, but they may represent attractive starting points for optimisation in the LO phase.

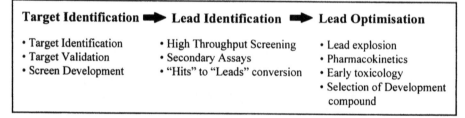

FIGURE 1: The Drug Discovery Process

TARGET IDENTIFICATION AND SELECTION

A microbial genome sequence provides a genetic blueprint of the molecular components that constitute a functional cell. By itself, however, it does not identify directly the subset of genes that encode antimicrobial targets - the challenge in antimicrobial target discovery, therefore, is to identify the subset of genes that represent attractive targets. The clinical profile of a new agent can be used to focus initial target identification research. Defined attributes within the clinical profile can be identified and translated into features required of a new molecular target. The example described in Figure 2 is for a broad spectrum agent (antibacterial or antifungal) with other attributes consistent with it being a safe and effective therapy.

The features required of the molecular target can be used to derive selection criteria, equivalent to filters, that can be applied to interrogate genomic databases sequentially. The example described above can be initially addressed by four filters: spectrum, selectivity, function and essentiality. In effect, this analysis represents a decision tree in which targets that meet one criteria are then filtered against further criteria. The objective is to process a large set of potential targets rapidly, i.e. the total repertoire of

134

Clinical Profile	Molecular Target
• Broad Spectrum $\longrightarrow$	• Present in all Target Species
• Cidal/static growth arrest $\longrightarrow$	• Essential Function
• Active versus resistant strains $\longrightarrow$	• Novel Target, Novel Compound
• No/low side effects $\longrightarrow$	• No host functional counterpart

FIGURE 2: Translation of a clinical profile into the requirements of a molecular target

genes within a genome, to a smaller number that warrant investigation in much greater detail. Each filter, therefore, can be thought of as a "module" that can be inserted into a bioanalysis pipeline at any position, depending upon the requirements of the project (Fig. 3).

To a large extent, the order in which these filters are applied is dependent upon the information and resources available at any particular time. For example, in the bacterial arena there are many bacterial genome sequences available via public or proprietary databases (see below) that allow the spectrum of a molecular target to be evaluated at an early stage in the bioanalysis pipeline. In contrast, the number of fungal genome sequences is restricted and, the same analysis of spectrum in the identification of new antifungal targets is better placed towards the end of the bioanalysis pipeline, and in the near future may involve "wet biology" until more genome sequences become available. From the outset, therefore, it should be recognised that the data sets for inclusion into this analysis will change over time in terms of quantity and quality.

Evaluating the Spectrum of a Molecular Target by *In-silico* Sequence Comparisons

The spectrum of the molecular target is critical since it should be present in all the organisms against which a drug needs to be active. Using sequence comparison algorithms, such as BLAST or FASTA, it is possible to establish whether a particular gene sequence and, therefore, gene function is present within a organism. As the number of genome sequences increases to include all relevant bacterial and fungal pathogens the analysis of spectrum will potentially be delivered completely by *in-silico* methods.

The first micro-organism to have its genome completely sequenced was the Gram negative bacterium, *Haemophilus influenzae* (Fleischmann *et al.*, 1995). In the two years or so since this landmark publication a further 10 genome sequences have been reported including key model organisms and pathogens such as *Eschericha coli* (Blattner *et al.*, 1997), *Bacillus subtilis* (Kunst *et al.*, 1997), and *Helicobacter pylori* (Tomb *et al.*, 1997). *Saccharomyces cerevisiae* is the only fungal species to have had its genome sequence completely elucidated (Goffeau *et al.*, 1996), although other programmes focused on key pathogens such as *Candida albicans* and *Aspergillus fumigatus* are underway. That many more bacterial than fungal genomes have been determined probably reflects that these genomes are smaller in size and are less complex in terms of overall organisation. A useful overview of these projects plus on-going microbial genome sequencing projects can be obtained by reference to the Microbial Genome Database at The Institute of Genomic Research (www.tigr.org). In addition to these initiatives a number of commercial vendors, in particular, Incyte Pharmaceuticals Inc.(www.incyte.com) and Genome Therapeutics Corporation (www.genomecorp.com)

A. Bioanalysis Pipeline for Antifungal Target Identification

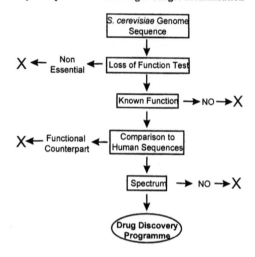

B. Bioanalysis Pipeline for Antibacterial Target Identification

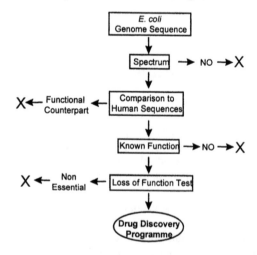

FIGURE 3: Schematic representation of bioanalysis pipelines to identify new antifungal and antibacterial targets. The order in which the different selection modules are applied differs in the two pipelines and is determined by the information available at the time of analysis. See text for details.

136

Have microbial genome sequencing programmes that are available to industrial partners on payment of database subscription fees.

For analysis of spectrum, organisms need to be grouped together in a way that is relevant to the clinical profile of the drug that is being targeted. This grouping can be in terms of clinical endpoint, e.g. to be effective in respiratory tract infections of bacterial origin, the target should be conserved in at least, *Staphylococcus aureus*, *Pseudomonas aeruginosa*, *Klebsiella pneumoniae* and *Streptococcus pneumoniae*, or by some other criteria that provide an effective filter, e.g. Gram positive versus Gram negative bacteria, or budding yeast versus filamentous fungi. If microbial genome sequence databases are configured appropriately, such groupings simply reflect alternative cuts across the database that can be performed rapidly by non expert users. The capability to perform comparative genomics on this scale will be a major bioinformatic focus in the years to come.

Identification of Microbial Specific Targets by *In-silico* Sequence Comparisons

Ideally, the molecular targets that are chosen for progression into drug discovery programmes should not be present, or should be substantially different, in the host. Focusing on this class of targets contributes to reducing the risk of failure, as it offers an opportunity for identifying new agents that are not impeded by mechanism-based toxicity. This does not mean, however, that compounds identified against these targets will be free from toxicity due to interference with other mechanisms via secondary and tertiary modes of action. This issue, however, can be addressed for those compounds that have been identified in the screening operation and can be used to kill projects early in the research process.

Comparative genomic analysis can be used to prioritise molecular targets on the grounds of selectivity. In this analysis the gene and/or protein sequence of each molecular target is compared, using algorithms, such as BLAST and FASTA, to human sequences in the public domain and proprietary databases. Although the complete sequence of the human genome will not be available for the next few years, there has been a substantial effort to sequence all of the expressed genes in the form of Expressed Sequence Tags (ESTs). The dbEST database (http://www.ncbi.nlm.nih.gov/dbEST/) currently contains over 800,000 human ESTs and proprietary databases, such as the LifeSeq^{TM} database from Incyte Pharmaceuticals Inc has in excess of 2,500,000 ESTs. Together these probably represent the majority of the genes that will be uncovered in the human genome. This information, therefore, provides an opportunity to select antimicrobial targets that have no obvious human counterpart. However, this analysis will require regular updates as the human genome project continues, and it will need to incorporate new analysis tools as our ability to predict secondary and tertiary structure, and function, from DNA sequence information improves.

Identification of Functions Essential for Cell Viability

Antimicrobial target discovery within ZENECA is largely focused on genes that are essential for cell viability. Drug discovery programmes based around these targets have the potential to deliver agents with a cidal mode of action that can be tested using conventional *in-vitro* sensitivity methodology. The definition of *in-vitro* essentiality is an operational one that will change over time as our knowledge base increases. Ultimately, *in vitro* conditions must be employed in molecular genetic studies that more accurately reflect the conditions encountered by the pathogen in the host. Some molecular genetic protocols such as Signature Tagged Mutagenesis (Hensel *et al.*, 1995) and *In Vivo* Expression Technology (Slauch *et al.*, 1994) have been developed to specifically identify the subset of genes that are required for the development and maintenance of infections in the host These factors are often referred to as virulence or pathogen maintenance factors. It is not clear what role this class of target will play in mainstream antimicrobial drug discovery as they have the potential to deliver agents that are effective against the micro-organism when it is in the host but are inactive under *in vitro* conditions. Such a profile would increase the technical challenges associated with the drug discovery programme (and thereby increase the risk of failure) and would also require changes in clinical practice to gain widespread acceptance.

Examination of completed genome sequences has revealed that approximately 50-60% of the genes identified by genome sequencing are novel, i.e. there existence was only established in the sequencing project. This was surprising for well worked model organisms such as *E. coli* and *S .cerevisiae*, but may be even higher for many pathogens that have not previously been extensively studied. As a consequence, a large number of the genes that have been identified require some form of genetic validation prior to progression into drug discovery programmes. Numerous techniques have been described that can be used for the identification of essential genes. High throughput gene knockout protocols are available for some model organisms including *S. cerevisiae* (Wach *et al.*, 1994) and *E.coli* (Russell *et al.*, 1989), that allow a large number of genes to be tested for essentiality. Equivalent technologies for clinically important pathogens are under development but are not yet as user friendly. In practice, therefore, organisms such as *S. cerevisiae* and *E. coli* function as a primary filter in the identification of essential genes that may then be further validated by genetic studies in other pathogens.

Understanding Gene Function Facilitates the Development of Drug Discovery Programmes

To develop a drug discovery programme around a molecular target, sufficient information about the function of that target must be available to guide the development of compound screening assays. Moreover, the format of the assays that are developed must be compatible with screening thousands of compounds at a time if the drug discovery programme is to successfully exploit high throughput screening and identify new agents. Most pharmaceutical companies appear to prefer targets where there is some successful precedent such as enzyme targets, rather than protein/DNA or protein/protein interactions. Further priority based on the feasibility of obtaining active compounds may be given to specific families of targets, e.g. proteases, kinases, to match specific in-house expertise. Examination of the genome sequences of *S. cerevisiae* and *E.coli*, however, has revealed that a biological/biochemical function cannot be ascribed to 50 -60% of the genes found within these organisms (Goffeau *et al.*, 1996, Blattner *et al.*, 1997). Although this subset of genes, referred to as orphans, will no doubt contain representatives that are essential for cell viability they will remain largely intractable for drug discovery until some function can be ascribed. For *S. cerevisiae*, the research community has initiated a large scale functional analysis project, the European Functional Analysis Project (EUROFAN), to study orphan genes. This is a collaborative project between 140 academic groups with the primary objective of assigning biological functions to approximately 1000 orphan genes. To date, no equivalent consortium has been established in the bacterial arena, which is particularly surprising for *E. coli* given the position it occupies in biological research.

APPLICATION OF GENOMICS IN THE LEAD IDENTIFICATION AND OPTIMISATION PHASESE OF DRUG DISCOVERY

The rapid acquisition of genome sequences for many bacterial and fungal pathogens has focused attention on the key technologies, and their applications, that will impact in the post genome era. Analysis of gene expression is one particular area of importance, where efforts are underway to systematically document the expression of every gene within a genome under different environmental conditions. Such studies have traditionally monitored the expression of small numbers of genes using techniques such as Northern blot or RT-PCR to evaluate RNA levels, or Western blotting to determine protein levels. The genomics era has catalysed the development of technologies, such as Differential Display PCR (Liang and Pardee, 1992), Serial Analysis of Gene Expression (SAGE, Velculescu *et al.*, 1995), and microarrays of DNA fragments (Schena *et al.*, 1995) or oligonucleotides (Lockhart *et al.*, 1996), that permit the expression of many genes to be studied simultaneously. In human biology, these protocols are being used to determine levels of gene patterns in various organs, tissues, and cells from "healthy" and "diseased" samples in an attempt to identify new molecular targets for therapeutic intervention.

Some of these techniques, in particular SAGE (Velculescu *et al.*, 1997), oligonucleotide microarrays (Wodicka *et al.*, 1997), and DNA fragment microarrays (Lashkari *et al.*, 1997) have been used to study gene expression in *S. cerevisiae*. In fact, this organism is often exploited as a test bed for the development of these technologies. Preliminary studies at ZENECA have evaluated the potential of the latter technique as a tool for elucidating the mode of action of antimicrobial compounds. This study focused on the sterol biosynthetic pathway in *S .cerevisiae* (Fig. 4), and exploited a strain of yeast in which the chromosomal copy of an essential gene in the sterol biosynthetic pathway, *ERG7* (Corey *et al.*, 1994, Shi *et al.*, 1994), encoding oxidosqualene cyclase had been knocked out. This strain was kept alive by a plasmid copy of an *ERG7* gene under the control of the *GAL10* promoter (Johnston and Davis, 1984); the strain, therefore, is viable when grown on galactose medium but dies when shifted to glucose medium. When the strain is grown on galactose medium the *GAL* genes, required for utilisation of galactose as a carbon source are expressed highly (Johnston, 1987) and the *ERG* genes, required for synthesis of sterols (Barratt-Bee and Dixon, 1995), are also expressed (Fig. 4B). On shift to glucose medium, however, the *GAL* genes are repressed and, consequently, the *ERG* genes are induced due to depletion of oxidosqualene cyclase and blockade of sterol biosynthesis. A very similar pattern of *ERG* gene regulation is observed when wild type cells are exposed to terbinafine, an inhibitor of sterol biosynthesis that blocks the pathway at an adjacent step to *ERG7* (not shown). By comparing the gene expression "fingerprints" of gene knockout strains and compound treated cells this technology, therefore, may be applicable for the confirmation, or determination, of the mode of action of novel antimicrobial agents. If successful, it will impact on the oldest method of drug hunting, of screening for antimicrobial activity, and

A.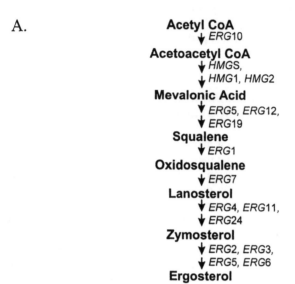

Acetyl CoA
↓ *ERG10*
Acetoacetyl CoA
↓ *HMGS,*
↓ *HMG1, HMG2*
Mevalonic Acid
↓ *ERG5, ERG12,*
↓ *ERG19*
Squalene
↓ *ERG1*
Oxidosqualene
↓ *ERG7*
Lanosterol
↓ *ERG4, ERG11,*
↓ *ERG24*
Zymosterol
↓ *ERG2, ERG3,*
↓ *ERG5, ERG6*
Ergosterol

B.

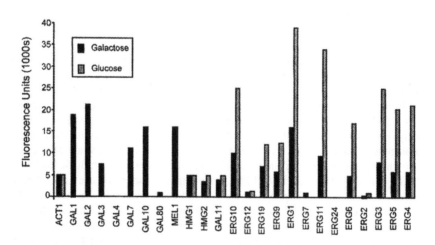

FIGURE 4: A. Schematic representation of the major steps in the sterol biosynthetic pathway (Barratt-Bee and Dixon, 1995). For the gene expression studies a strain of *S. cerevisiae* was used in which the chromosomal *ERG7* gene (encoding oxidosqualene cyclase) had been knocked out and which carried a complementing plasmid with an *ERG7* gene under the control of the *GAL10* promoter. B. Results of the gene expression analysis. The above strain was grown in medium that contained either galactose or glucose as a carbon source. Poly A+ RNA was extracted and used to probe microarrays using protocols similar to those described in Lashkari *et al.*, 1997. Fluoresence read outs were normalised with respect to the levels of actin gene (*ACT1*) expression.

potentially turn a strategy based on empirical observation into a molecular based rational design process.

Similar approaches also have potential applications in mammalian cells, which may provide the basis for high throughput screening for molecular toxicity. By comparison to similar data from gene expression microarrays of experimental animals, the first links between organ toxicity and molecular processes could be made. The laboratory of the future is likely to contain data banks of this type of information, alerting the experimenter to toxicological issues at an early stage, long before the compound is exposed to an animal.

CONCLUSIONS

The future of anti-infective research has never been more exciting. The framework of information provided by genomics is enabling the application of novel approaches across the drug discovery process. In target selection various pharmaceutical companies are attempting to exploit this new technology, and employing different strategies. The basic issues interrogated by these differing approaches, of essentiality, spectrum, potential selectivity, and functionality are inevitably the same. The major differences occur in the order and scale in which these questions are addressed. Unlike the historical situation, we now have the total genome of micro-organisms from which to select targets. The choice has never been better!

Beyond target selection, technology developed primarily to exploit genomic information has the potential to dramatically influence the drug-hunting process, characterising the effects of compounds and genes in a similar format. The compound collections of the future may include expression profiles, so that the effects of a compound can be matched *in silico* to the effects of genetically impairing a target protein. Biochemical screening may no longer be essential in high throughput mode, only being required to confirm the prediction.

The problems that are being encountered in the clinic with contemporary antimicrobial therapies are demanding a range of new anti-microbial agents with novel modes of action. Microbial genomics is providing the information and the tool kit to make this happen, and at the same time facilitating a more rational approach to target selection based on clinically relevant properties.

REFERENCES

ASM (1995) Report of the ASM Task Force on Antibiotic Resistance. *Supplement to Antimicrobial Agents and Chemotherapy*

Barratt-Bee, K. and Dixon, G.K. (1995). Ergosterol Biosynthesis Inhibition: A Target for Antifungal Agents. *Acta Biochemica Polonica* **42**:465-480.

Blattner, F.R., Plunkett III, G.., Bloch, C.A., Perna, N.T., Burland, V., Riley M., Collado-Vides, J., Glasner, J.D., Rode, C.K., Mayhew, G.F., Gregor J., Davis, N.W.,

Kirkpatrick, H.A., Goeden, M.A., Ros,e D.J., Mau, B., and Shao, Y. (1997). The Complete Genome Sequence of Eschericha coli K12. *Science* **277**: 1453-1462.

Chopra, I., Hodgson, J., Metcalf, B. and Poste, G. (1996) New approaches to the control of infections caused by antibiotic-resistant bacteria. *Journal of Antimicrobial Agents,* **275**: 401-402.

Corey, E.J., Matsuda S.P. and Bartel, B. (1994). Molecular Cloning, Characterization, and Overexpression of *ERG7*, the *Saccharomyces cerevisiae* Gene Encoding Lanostrol Synthase. *Proceedings of the National Academy of Sciences, USA* **91**:2211-2215.

Davies, J.(1994) Inactivation of antibiotics and dissemination of resistance genes. *Science* **264** 375-382.

Fleischmann, R.D., Adams, M.D., White, O., Clayton, R.A., Kirkness, E.F., Kerlavage, A.R., Bult, C.J., Tomb, J.F., Dougherty, B.A., Merrick, J.M., *et al.,* (1995). Whole-genome random sequencing and assembly of *Haemophilus influenzae* Rd. *Science* **269**: 496-512.

Goffeau, A., Barrell, B.G., Bussey, H., Davis, R.W., Dujon, B., Feldmann, H., Galibert, F., Hoheisal, J.D., Jacq, C., Johnston, M., *et al.,* (1996). Life With 6000 Genes. Science **274**: 546-567.

Gold, H.S. and Moellering, R.C.(1996) Antimicrobial-drug resistance. *New England Journal of Medicine,* **335,** (19) 1445-1453.

Hensel, M., Shea, J.E., Gleeson, C., Jones, M.D., Dalton, E. and Holden, D.W.(1995). Simultaneous Identification of Bacterial Virulence Genes by Negative Selection. *Science* **269** 400-403.

Johnston, M. and Davis, R.W. (1984). Sequences That Regulate the Divergent *GAL1-GAL*10 Promoter in *Saccharomyces cerevisiae. Mol. Cell. Biol.* **4**:1440-1448.

Johnston, M. (1987). A Model Fungal Regulatory Mechanism: The GAL Genes of *Saccharomyces cerevisiae. Microbiological Reviews* **51**:458-476.

Kunst, F., Ogasawara, N., Moszer, I., Albetini, A.M., Alloni, G., Azevedo, V., Bertero, M.G., Bessieres, P., Bolotin, A., et al. (1997). The Complete Genome Sequence of the Gram Positive Bacterium *Bacillus subtilis. Nature* **390**:249-256.

Lashkari, D.A., DeRisi, J.L., McCusker, J.H., Namath, A.F., Gentile, C., Hwang, S.Y., Brown P.O. and Davis, R.W.(1997). Yeast Microarrays for Genome Wide Parallel Genetic and Gene Expression Analysis. *Proc. Natl. Acad. Sci. USA* **94**:13057-13062.

Levy, S.B. (1992) The Antibiotic Paradox: How Miracle Drugs are Destroying the Miracle, New York & London Plenum Press.

Liang, P. and Pardee, A.B. (1992). Differential Display of Eukaryotic Messenger RNA by Means of the Polymerase Chain Reaction. *Science* **257**:967-971.

Lockhart, D.J., Dong H., Byrne, M.C., Follettie, M.T., Gallo, M.V., Chee, M.S., *et al.*, (1996). Expression Monitoring by Hybrisidisation to High-Density Oligonucleotide Arrays. *Nature Biotechnology* **14**:1675-1680.

Neu, H.C., (1992) The crisis in antibiotic resistance. *Science,* **257,** 1064-1073.

Rockerfeller University Workshop (1994) Multiple-antibiotic resistant pathogenic bacteria, *Special report - New England Journal of Medicine,* **330,** 1247-1251.

Russell, C.B., Thaler, D.S. and Dahlquist, F.W. (1989). Chromosomal Transformation of Eschericha coli recD Strains with Linearised Plasmids. *J. Bacteriology* **171**:2609-2613.

Schena, M., Shalon, D., Davis, R.W. and Brown, P.O. (1995). Quatitative Monitoring of Gene Expression Patterns with a Complimentary DNA Microarray. *Science* **270**: 467-470.

Shi, Z., Buntel, C.J. and Griffin, J.H. (1994). Isolation and Characterization of the Gene Encoding 2,3-Oxidosqualene-Lanosterol Cyclase from *Saccharomyces cerevisiae. Proceedings of the National Academy of Sciences, U.S.A.* **91**:7370-7374.

Slauch, J.M., Mahan, M.J., Mekalanos, J.J. (1994). *In vivo* Expression Technology for Selection of Bacterial Genes Specifically Induced in Host Tissues. *Methods in Enzymology* **235**: 481-92.

Tomb, J.F., White, O., Kerlavage, A.R., Clayton, R.A., Sutton, G.G., Fleischmann, R.D., Ketchum, K.A., Klenk, H.P., Gill, S., Dougherty, B.A., Nelson, K., Quackenbush, J., Zhou, L., Kirkness, E.F., Peterson, S., Loftus, B., Richardson, D., Dodson, R., Khalak, H.G., Glodek, A., McKenney, K., Fitzegerald, L.M., Lee, N., Adams, M.D. and Venter, J.C. (1997). The Complete Genome Sequence of the Gastric Pathogen *Helicobacter pylori. Nature* **388**:539-547.

Velculescu, V.E., Zhang, L., Vogelstein, B. and Kinzler, K.W. (1995) Serial Analysis of Gene Expression. *Science* **270**: 484-487.

Velculescu, V.E., Zhang, L., Zhou, W., Vogelstein, J., Basrai, M.A., Bassett, D.E., Hieter, P., Vogelstein, B. and Kinzler, K.W. (1997). Characterisation of the Yeast Transcriptome. *Cell* **88**: 243-251.

Wach, A., Brachat, A., Pohlmann, R. and Philippsen, P. (1994). New Heterologous Modules for Classical or PCR-based Gene Disruptions in *Saccharomyces cerevisiae. Yeast* **10**:1793-1808.

WHO (1996) The World Health Report, WHO Publications - Geneva.

Wodicka, L,. Dong, H., Mittmann, M., Ho, M. and Lockhart, D.J. (1997). Genome wide Expression Monitoring in *Saccharomyces cerevisiae. Nature Biotechnology* **15**:1359-1367.

PATENTING OF GENOME SEQUENCES: INVENTIONS OR MERE DISCOVERIES?

R. KEITH PERCY

BTG plc, Patents Division, 101 Newington Causeway, London, SE1 6BU, UK.

ABSTRACT

Patent law has long upheld the patentability of products isolated from, or the same as found in nature. DNA molecules from human DNA sequencing projects do not exist as such in nature, but the question of whether they are an unpatentable discovery is still relevant. Discovery is the finding of new properties or uses of a known thing or finding the existence of something not previously known to exist. Mere discoveries are unpatentable, but the isolated DNA, being free of its natural environment, is not a discovery. Whether or not it is inventive depends on whether its isolation was obvious and whether it has a sufficiently well-defined use. Some of the uses stated might well be regarded as adequate, as patent law does not require high standards for this purpose.

THE PATENT SYSTEM

The world has declared itself in favour of patents, as is evident from the fact that they are available in more than 150 countries, from Albania to Zimbabwe. The only countries which do not possess a patent system are small and isolated, examples being Andorra and Tonga. Furthermore, nowadays nearly all countries grant patents for chemicals, including DNA and proteins. Only a few countries do not grant patents for micro-organisms.

Whether the patent system is a good thing or not is irrelevant, in the sense that the world believes it to be. No country has ever left the patent system. Even the People's Republic of China has joined it and there are now few countries in the world which have not recognised that a healthy economy depends on the individual profit motive, in which patents play a part.

WHAT PATENTS DO

A patent is a limited monopoly, defined by the "claims" of the patent, which gives the owner the right to prevent others from making, selling, carrying out or using the invention commercially. Experimental use of the invention for research purposes is not barred by a patent. In most countries the monopoly is ordinarily for a term of 20 years.

A patent is also a publication of the invention. The rationale of the patent system is basically twofold. Firstly, it is a trade-off, by which the state grants the monopoly in return for the publication of a full description of the invention. This is judged to be better than the alternative of secrecy. Secondly, the patent system encourages research and development expenditure.

PHARMACEUTICAL PATENTS

It might be thought that patents simply give pharmaceutical companies permission to print money. In fact, this is not so, for a variety of reasons, chiefly the large sums which have to be expended on research and development and the ability of many governments to control prices of medicines sold through the state.

If this view were wrong, redress could be sought through the patent system, by shortening the term of the monopoly on pharmaceuticals. Far from this happening, however, the reverse has been the case. The time taken to develop a pharmaceutical, test it thoroughly and obtain approval from the regulatory authorities is long. An estimate in Europe is 12 years from filing the patent application (European Commission, 1990), but for biotechnological products it is probably even longer. To counter this problem, the patent system has evolved to extend the monopoly beyond 20 years to compensate for these delays. Extensions of up to 5 years are now obtainable in the major countries of the world for pharmaceutical patents, the precise extension corresponding approximately to the delay in obtaining regulatory approval.

OPPOSITION TO GENE PATENTS

"No patents on life" groups argue that all normal genes are owned collectively by mankind and are therefore incapable of yielding an exclusive right to any person. Leaving aside that patent rights do not confer ownership of the subject matter and accepting this argument as genuinely altruistic, it is questionable whether its use to attack the patent system is cleverly directed. These groups would surely agree that, in general, the development of medical treatment based on DNA, or the proteins which are encoded by it, is a good thing. If these genes and proteins are not protected in some way by an exclusive right, there will be no incentive for research and development. If genes are part of a commonly-owned, collective heritage, a remedy to the perceived injustice would be for the company marketing the treatment to pay a royalty to all mankind. Such royalties are well known: they are called taxes and whether the revenue thus raised would be passed on to mankind must be considered doubtful, in the light of experience of other specific taxes.

DISCOVERIES ARE UNPATENTABLE

Many patent laws state that discoveries are unpatentable. European countries' patent laws conform to the European Patent Convention which contains a specific prohibition (European Patent Convention, 1973, as amended 1991).

US patent law has no counterpart in the statute. Case law has resulted in some patent applications being rejected because the invention was claimed in such a generalised way that it was held not to define a process. An early example was Morse's invention of the telegraph which was claimed as follows (US Supreme Court Decision, 1854).

"Eighth. I do not propose to limit myself to the specific machinery or parts of machinery described in the foregoing specification and claims; the essence of my invention being the use of the motive power of the electric or galvanic current, which I call electromagnetism, however developed, for marking or printing intelligible characters, signs or letters, at any distances, being a new application of that power of which I claim to be the first inventor or discoverer."

The problem with Morse's claim was not so much that he had made a mere discovery, but that he had not defined his process sufficiently concretely.

In Japanese patent law mere discoveries are unpatentable if not "industrially applicable" (Japanese Patent Law, 1959-1994).

PRODUCTS OF NATURE

Modern patent laws allow the patenting of substances themselves. None of the major countries has a specific prohibition in statute law on patenting "products of nature". Perhaps this is not surprising. After all, a product which exists in nature is not new (in its natural form) and patent laws already require novelty.

Of course, many patents have been granted for products isolated from nature. The favourite old example is Pasteur's US patent (US Patent 141,072, 1873) which claimed "Yeast, free from organic germs of disease, as an article of manufacture". In more modern times there have been two "landmark" decisions upholding the patenting of products of nature. One is the vitamin B-12 case in the US, where the validity of the following claim was upheld:

"The compound vitamin B-12, an organic substance containing cobalt, together with carbon, nitrogen, hydrogen, oxygen, and phosphorus, said compound being a red crystalline substance soluble in water, methyl and ethyl alcohol and phenol, and insoluble in acetone, ether and chloroform, and exhibiting strong absorption maxima at about 2780 Angstroms, 3610 Angstroms and an LLD activity of about 11,000,000 LLD units per milligram." (US Patent 2,563,794, 1951)

This is the same compound as present in liver extracts.

The other important decision was of the German Federal Patent Court which dealt with a chemical compound, antamanide, a decapeptide (Decision of German Federal Patent Court, 1977). This substance, prepared synthetically, was apparently similar to a substance found in a naturally-occurring, poisonous fungus, which counteracted the fatal effect of the poison. The Patent Office alleged that the decapeptide must be present in the natural state in the fungus and was therefore unpatentable. The Federal Patent Court could not prove that the substance was in fact a natural substance, but decided the matter as if it was, upholding the claim. The Court emphasised that natural and synthetic compounds should not be treated differently.

In Great Britain there have been no notable decisions. The reason is that British Patent Law found a neat way of avoiding trouble when the Patents Act was revised in 1949. The Act provided:

> "Where a complete specification claims a new substance, the claim shall be construed as not extending to that substance when found in nature." (British Patents Act, 1949)

In other words, anyone who sells, uses etc. a natural product cannot infringe, because the claims, however drafted, cannot be construed to cover it. Most patent people consider this an excellent solution and regret that the new British Patents Act did not include this wording. The only problems that might arise are in a few rare cases where the product was not naturally occurring when the patent application was filed, but became so later. An example is the deliberate mutation of an influenza virus in the laboratory with the aim of creating a vaccine which would be effective when the starting virus mutated naturally in the wild through "antigenic drift" (UK Patent 1,443,958, 1976).

The precise form of claim acceptable to the Patent Office differs, notably in the US where it frequently has to begin: "An isolated and purified cDNA" or "A biologically pure culture". However, if the compound is new in its isolated form, unobvious and has a use, it is patentable as such.

GENE SEQUENCES

Much human gene sequencing is concentrated on cDNA. Since cDNA is merely the complement of mRNA, the "product of nature" issue does not arise. DNA isolated from the human genome, with its terminal phosphate and hydroxy groups unbound to adjacent nucleotides, does not exist in nature as such either. Nevertheless, the invention/discovery question is still pertinent. Such DNA sequences can be sub-divided into two cases:

- those for which mRNA or genomic DNA was known to exist, especially those which are translated into a known protein;

- those which mRNA or genomic DNA was not known to exist.

The first category has proved to be patentable in Europe, in relatively early cases only. For example, the patent on recombinantly-derived erythropoietin was upheld in an amended form, only after lengthy opposition proceedings at the European Patent Office in which more than 500 documents were cited (European Patent Office Appeal Board Decision T412/93, 1995). In other cases patents have been refused (European Patent Office Appeal Board Decisions T249/88, 1996 and T386/94, 1997). The problem is, that once one knows the amino acid sequence of the protein or has some way of monitoring the products of expression of the gene, it is relatively easy to select clones containing the gene by probing them with a pool of labelled oligonucleotides, or selecting expression library clones which give rise to a known phenotype. Thus, the work becomes obvious.

The same difficulty faced patent applicants in the US until recently, when case law came to their rescue rather unexpectedly. In the case of *In re Deuel*, the US Court of Appeals, Federal Circuit (1995), the Court, which is two layers of the judicial system up from the Patent Office, upheld a patent application for DNA encoding human heparin binding growth factor (HBGF) of 168 amino acids defined by its DNA sequence. The technical background is briefly as follows.

The inventors isolated and purified HBGF from bovine uterine tissue, found that it exhibited mitogenic activity, and determined the first 25 amino acids of the N-terminal sequence of the protein. They then isolated a cDNA molecule encoding bovine uterine HBGF by screening a bovine uterine cDNA library with an oligonucleotide probe designed using the N-terminal sequence of the bovine HBGF. The bovine HBGF cDNA was then isolated and a clone of this cDNA used as a probe to screen a human placental cDNA library. The inventors then purified and sequenced the human placental cDNA clone, which was found to consist of a sequence of 961 base pairs. The predicted human placental and bovine uterine HBGFs each have 168 amino acids, of which 163 are identical. The prior literature related to a group of heparin-binding brain mitogens having the same 19 N-terminal amino acids in human and bovine versions, so the examiner contended that it would have been obvious to use the amino acid sequence to construct a probe to isolate HBGF DNA from uterine tissue. The amino acid sequence was the same in these brain mitogens as in the uterine HBGFs although this was not known at the time. The applicants' argument, that one would not expect these sequences to be the same because of the different bodily tissues from which the HBGFs were derived, was not accepted by the Patent Office.

Instead of simply reversing the Patent Office decision by upholding that argument based on tissue types, the US Court of Appeals held that the claim to the DNA defined by the nucleotide sequence was patentable, because of the redundancy of the genetic code. The decision appears to mean that there is invention in designing an oligonucleotide probe, based on known amino acid sequence and coding usage principles, to select cDNA clones, followed by DNA sequencing of the clone. Molecular biologists know that, except in rare cases, it is absurd to think of this as other than routine.

The position of Japanese patent law in similar circumstances is not fully clear, but the Japanese Patent Office requires a showing that the DNA coding for the protein has "remarkable advantageous effects over other genes having a different base sequence encoding the protein" (Japanese Patent Office, 1997). (The guideline has thus been drafted in a manner more appropriate to the patentability of DNA which has been mutated by human intervention.)

Turning to the second case, if a substance is not known to exist in nature and has been isolated for the first time, it is novel and likely to be unobvious and therefore, so long as it is useful, patentable.

SCREENING PROGRAMMES

The most controversial aspects of the patenting of genomic nucleic acid are the so-called expression sequence tags and the genes of which they form a part. These can be considered a special case, because they are the products of an enormous screening programme. New compounds obtained by screening thousands of cultures of naturally-occurring micro-organisms for anti-cancer, anti-viral compounds etc. are normally patentable. Human DNA cloning and sequencing can be argued to differ. Firstly, the chances of success can be enhanced by starting from a cDNA library of a relevant tissue. Secondly, there is some possibility of deducing the properties of the encoded protein by searching for similar sequences of known proteins in a sequence bank. Is this discovery, not invention? What exactly is "discovery" anyway?

DEFINITIONS OF "DISCOVERY"

Dictionary definitions, inevitably a starting point, are of limited help. The relevant meaning of the noun in the Oxford English Dictionary (1989) is "The finding out or bringing to light of that which was previously unknown; making known". In Chambers' 21st Dictionary (1996), the noun is defined by reference to the verb as "1. to be the first person to find something or someone; 2. to find by chance, especially for the first time; 3. to learn of or become aware of for the first time".

A definition more closely related to science would be helpful. Intriguingly, there is a Geneva Treaty on the International Recording of Scientific Discoveries (1978) and this defines the term "scientific discovery" as "the recognition of phenomena, properties or laws of the material universe not hitherto recognised and capable of verification". This is a definition for the particular purpose of the Treaty, never ratified, which was clearly intended not to conflict with inventions arising from the practical application of these phenomena, properties and laws.

Reid (1984) says of the exclusion of discoveries from patentability:

> "The discovery" exclusion is rather more difficult to grasp. What it means essentially is that there must be some kind of tangible embodiment to hand before a patent is obtainable and that mere discovery of new knowledge - say, of some new physical or chemical effect - is not as such protectable.",

while Phillips & Firth (1995) add:

> "A discovery - therefore merely to find a hitherto unknown substance which exists in nature - is not to make an invention; nor is it an invention to discover a physical property of such a substance (e.g. that it is magnetic, water-soluble or has good heat conduction properties). However, a discovery may lead to a product or process which is patentable."

In connection with the discovery that an old thing has new properties, the celebrated British patent barrister Blanco White (1974) said:

> "An extreme case of a specification claiming nothing that is "new or alleged to be new" is found where the article or process claimed is old:

the alleged invention consisting merely in the discovery that an old article or process is useful for a particular new purpose, and the alleged distinction in the claims from what is old consisting merely in an indication that the article or process is claimed "for" that purpose. Such claims are invalid: a thing is not made a different thing by a statement that is to be used for a particular purpose or with a particular end in view.
Patents are not granted for the mere discovery that a particular known thing has hitherto unknown properties ..."

He meant here that the words "for the purpose of ..." do not limit the patented device, substance, etc., a principle long established in British patent law [Adhesive Dry Mounting v. Trapp (1910)]. In fact, this principle has been diluted by modern European patent law, where these words do limit the patented medical substance to the field of medical use. Leaving this quirky exception aside, however, it is still the law that if the novelty of a substance lies in its purpose, then one must claim the purpose, not the substance itself.

Summing up, discovery is the finding of new properties or uses of a known thing or finding the existence of something not previously known to exist.

DIFFERENTIATING BETWEEN INVENTION AND DISCOVERY

The subtle differences relevant here between a patentable invention and an unpatentable discovery are, therefore, as follows:

1. The finding of (say) an alkaloid in plant leaves or a gene in human chromosome 6 is a discovery. This means that the alkaloid when present in plant leaves is inherently unpatentable. Expressed another way, the leaves of a naturally growing plant are not patentable even if the plant grows only on an island never previously inhabited. Whether or not they have been found to contain an alkaloid is irrelevant. Similarly, DNA when present in a human chromosome is not patentable.

2. On the other hand, the alkaloid when extracted from the plant and the DNA excised from the chromosome are not discoveries. This is because their isolated form was not previously known to exist. Just because they are not discoveries does not mean that they are inventions. To be inventions, they must also be unobvious and useful.

In principle, screening could affect whether the DNA was obvious. In most cases, being the product of a screening programme will not make it obvious, because the amount of selection required to arrive at it will be rather high. To find that one particular optical isomer is responsible for the activity of a racemate (an equal mixture of two isomers) has been held uninventive (European Patent Office Appeal Board Decision T296/97, 1990). This is an extreme case: selecting one possibility out of six could well be inventive. Thus, for example, when searching for a cancer gene by comparing the genetic make-up of normal individuals and cancer patients, the number of gene differences is likely to be far in excess of six. The problem is to decide which of these are due to cause and not effect, and which are genuinely connected with the cancer.

The real nub of the matter is usefulness (utility). Patent law requires that the claimed subject matter must be useful or "industrially applicable". Although not all laws say so, it is generally accepted that at least one use must be stated in the specification when the patent application is filed. How do the human DNA sequencing patent applications stand up to this criterion?

THE CURRENT STATUS OF HUMAN DNA SEQUENCING PATENT APPLICATIONS

The earliest human DNA sequencing patent applications were those of the US National Institutes of Health (NIH) and the UK Medical Research Council (MRC). Both have been abandoned or withdrawn. The NIH application claimed several hundred nucleotide sequences from a brain cDNA library and was the only application to be seriously examined. The official letter from the Patent Office was made public. The application was objected to as not giving a clearly enough defined utility and as obvious over a prior art reference, mainly because it included all oligonucleotides of at least 15 nt within each sequence (Unpublished US Patent Application 07/837195, 1993). The MRC application claimed 1193 cDNA sequences from human brain and adrenal mRNA (Patent Co-operation Treaty Application, 1994).

A press report suggesting that the US Patent Office is about to grant patents on expression sequence tag cases is greatly premature (O'Brien, 1997). In fact, "restriction requirements" (objections that the patent application contains more than one invention and must be limited to one only) have been issued in 50 cases. Only after they have been answered does proper examination begin. In the European Patent Office, the proper examination of human DNA sequencing applications has not yet begun. It will be a long time before an authoritative body of case law in any country has built up.

Whether many patent applicants will claim expression sequence tags in the sense of pieces of DNA which fall short of the length of a gene, or contain intron sequence, is doubtful. The US Patent Office appears likely to refuse them unless they are claimed as "DNA consisting of ...", thus excluding the full length gene.

One of the prominent patent applicants in this field is Incyte Pharmaceuticals, Inc. of Palo Alto, California. Inspection of a sample of their patent applications suggest that they are not claiming arbitrary lengths of DNA, but fully sequenced genes. The manner in which Incyte describe their utility can be exemplified by their published PCT application relating to a gene obtained from a human pancreatic cDNA library (Patent Co-operation Treaty, 1996a):

> "The present invention provides a nucleotide sequence identified in Incyte 222689, uniquely identifying a new Pancreas-Derived Serpin (PDS) of the cystein protease family which was expressed in pancreatic cells. Because the nucleic acids specifically antibodies to PDS are useful in diagnostic assays for physiologic or pathologic problems of the pancreas. Increased expression of proteases are [*sic*] known to lead to tissue damage or destruction; therefore, a diagnostic test for the presence

and expression of PDS can accelerate diagnosis and proper treatment of such problems.

The nucleotide sequence encoding PDS has numerous applications in techniques known to those skilled in the art of molecular biology. These techniques include use as hybridization probes, use in the construction of oligomers for PCR, use for chromosome and gene mapping, use in the recombinant production of PDS, and use in generation of anti-sense DNA or RNA, their chemical analogs and the like."

Other Incyte patent applications are in a similar vein; see, for example, those relating to

- "a novel human tyrosine phosphatase .... which .... likely functions in the control of cellular proliferation, particularly cellular proliferation in tumour cell lines, in mitogen-stimulated cells and during tissue regeneration" (Patent Co-operation Treaty Application, 1997a).

- "a human EDG-2 receptor [*HEDG*] homolog"; Aan assay for abnormal expression of HEDG is a viable diagnostic tool for assessing the extent that RA has progressed" (Patent Co-operation Treaty Application, 1997b).

- "a novel human C5a-like receptor [*CALR*] homolog"; "an assay for unregulated expression of CALR can accelerate diagnosis and proper treatment of conditions caused by abnormal signal transduction events due to anaphylactic or hypersensitive responses ....." (Patent Co-operation Treaty Application, 1996b).

- "a novel human ICE homolog [*ICEY*] of the cysteine protease family". "Because ICEY is expressed in activated monocytes, the nucleic acids ...., polypeptides .... and antibodies to ICEY are useful in diagnostic assays based on the identification of ICEY associated with inflammation or inflammatory disease processes"(Patent Co-operation Treaty Application, 1996c).

The critical issue is whether the properties of the protein predicted or expressed by the DNA have been stated in a sufficiently precise way. In a recent letter to the US Patent Office, the US National Institutes of Health have pointed out, correctly, the analogy with traditional chemistry. A chemical intermediate having no other use than to make the final product depends for its utility on the utility of the final product. Similarly, a gene depends for its utility on the utility of the protein. The uses as probes, in assays and for chromosome mapping all depend on the protein having a biological function sufficiently well defined to make any of the uses real.

In other respects, the Incyte specifications are well drafted, containing pages of description of conventional techniques for hybridisation assays, chromosome mapping, raising of antibodies etc. The utilities stated do not seem manifestly implausible. Thus, there is a high probability that the US Patent Office will consider that they meet the requirements. The problem here, of course, is that if too strict a standard is applied to human DNA sequencing applications, it could affect many other inventions in other fields as well. Applicants are not normally required to provide any proof of asserted utility unless there is reason to doubt it.

All the same, they are just the kind of asserted utilities criticised publicly in the HUGO Statement on Patenting DNA Sequences (Caskey *et al.*, 1995). (HUGO is concerned that claims to the DNA and proteins themselves will dominate later-discovered uses. That s to say, the use cannot be put into practice without infringing the patents on the DNA and proteins themselves.)

The European Patent Office examiners look set to take a much tougher stance initially, probably along the lines that it is obvious to clone the human genome or cDNA, search for similarities in a database and make statements about industrial applicability based on the origin of the DNA library used and the uses of the protein of greatest homology found in the database. They may also object that the industrial applicability is not closely enough defined. However, patent applicants can appeal against the decision of the European Patent Office examiners. Case law is rarely authoritative at any lower level than the Appeal Boards. There is every prospect that the Appeal Boards will view the matter as one of unobvious selection and will not require a high standard of industrial applicability.

In short, patent applications of the kind cited above, where there is some description of biological function, even if not the most precise, might well be allowed. If Patent Offices challenge the stated uses, applicants can justify them by filing evidence. Therefore, much could depend on whether what is stated in the patent application as a use is in fact later realised.

REFERENCES

Adhesive Dry Mounting v. Trapp, (1910) *Reports of Patent Cases*, **27**, 341-353.

Blanco White, T.A. (1974) *Patents for inventions* (Fourth Edition), London: Stevens & Sons, section 1-208.

British Patents Act (1949), Section 4(7).

Caskey, C.T., Eisenberg, R.S., Lander, E.S., Staus, J. (March 15, 1995) *HUGO Statement on Patenting of DNA Sequences*, Bethesda, Maryland, USA: HUGO Americas.

Chambers' 21st Century Dictionary (1996), Edinburgh: Chambers.

Decision of the German Federal Patent Court 16 W(pat) 64/75 of 28 July 1977, "Antamanide", (1979) *International Review of Industrial Property and Copyright Law*, **10**, 494-500.

European Commission (1990) *Proposal for a Council Directive concerning the creation of a Supplementary Protection Certificate for Medical Products*, European Commission, 32 pp.

European Patent Convention (1973, amended 1978, 1994), Article 52(2).

European Patent Convention (1973, amended 1978, 1994), Article 52(4).

European Patent Office Appeal Board Decision T296/87, Enantiomers/HOECHST, (1990) *Official Journal of the European Patent Office*, 195-223; (1990) *European Patent Office Reports*, 337-351.

European Patent Office Appeal Board Decision T249/88 Milk Production/MONSANTO, (1996) *European Patent Office Reports*, 29-36.

European Patent Office Appeal Board Decision T412/93, Erythropoietin/KIRIN-AMGEN, (1995) *European Patent Office Reports*, 629-687.

European Patent Office Appeal Board Decision T386/94, Chymosin/UNILEVER, (1997) *European Patent Office Reports*, 184-200.

Geneva Treaty on the International Recording of Scientific Discoveries, Article 1(1)(i), reproduced in (1978) *Industrial Property Laws and Treaties*, Geneva: World Intellectual Property Organization, **VIII**, "Multilateral Treaties", Text 1-003; (April 1978) *Industrial Property*, Laws and Treaties section, Geneva: World Intellectual Property Organisation.

Japanese Patent Law N1 121 of 1959, Law N1 140 of 1962, Law N1 161 of 1962, Law N1 148 of 1964, Law N1 81 of 1965, Law N1 98 of 1966, Law N1 111 of 1966, Law N1 91 of 1970, Law N1 42 of 1971, Law N1, 96 of 1971, Law N1 10 of 1973, Law N1 46 of 1975, Law N1 27 of 1978, Law N1 30 of 1978, Law N1 45 of 1981, Law N1 83 of 1982, Law N1 78 of 1983, Law N1 23 of 1984, Law N1 24 of 1984, Law N1 41 of 1985, Law N1 27 of 1987, Law N1 91 of 1988, Law N1 30 of 1990, Law N1 26 of 1993, Law N1 89 of 1993 and Law N1 116 of 1994, Section 29(1).

New Guideline for examination in the field of bio-related inventions, Japanese Patent Office (March 1997), translation by Kawaguti & Partners, Tokyo, adapted from the official translation.

O'Brien, C. (1997) US Decision will not limit gene patents, *Nature*, **385**, 755.

The Oxford English Dictionary, Second Edition (1989), Oxford : Clarendon Press.

Patent Co-operation Treaty Application PCT/GB93/01467 (1994), Publication N1 WO 94/01548.

Patent Co-operation Treaty Application PCT/US96/06137 (1996a), Publication N1 WO 96/34957.

Patent Co-operation Treaty Application PCT/US96/08596 (1996b), Publication N1 WO 96/39511.

Patent Co-operation Treaty Application PCT/US96/08150 (1996c), Publication N1 WO 96/38569.

Patent Co-operation Treaty Application PCT/US96/12665 (1997a), Publication N1 WO 97/06262.

Patent Co-operation Treaty Application PCT/US96/10618 (1997b), Publication N1 WO 97/00952.

Phillips, J. and Firth, A. (1995) *Introduction to Intellectual Property Law* (Third Edition), London: Butterworths, section 4.11, p39.

Reid, B.C. (1984) *A practical guide to patent law*, Oxford: ESC Publishing Limited, p18.

UK Patent 1,443,958 (1976) granted under the 1949 Act.

Unpublished US Patent Application 07/837,195; Murashige, K. (October 1993), The NIH gene application's fate at the USPTO, *Patent World*, 15-18.

US Court of Appeals, Federal Circuit decision *In re Deuel*, (1995) *US Patent Quarterly*, **34** (*2nd series*), 1210-1216.

US Patent 141,072 (1873).

US Patent 2,563,794 (1951); US Court of Appeals, 4th Circuit Decision, Merck & Co Inc. v. Olin Mathieson Chemical Corp., (1958) *US Patent Quarterly*, **116**, 484-491.

US Supreme Court Decision O'Reilly v. Morse, (1854) US Reporter, **56**, 62-137.

APPLICATIONS IN MEDICINE

USING GENOMICS TO UNDERSTAND MICROBIAL PATHOGENESIS

E.R. MOXON, D. HOOD, M. DEADMAN, T. ALLEN

Molecular Infectious Diseases Group, Oxford University Department of Paediatrics, Institute of Molecular Medicine, John Radcliffe Hospital, Oxford OX3 9DU, UK

A. MARTIN, J.R. BRISSON, J.C. RICHARDS

Institute for Biological Sciences, National Research Council of Canada, Ottawa, Ontario, Canada, K1A 0R6

R. FLEISCHMANN, C. VENTER

The Institute for Genomic Research, 9712 Medical Center Drive, Rockville, MD 20850, USA

ABSTRACT

The availability of whole genome sequences of bacteria has brought about fundamental changes in microbiology and is paving the way for novel applications of this information that will benefit epidemiology, pathobiology, diagnosis and prevention of infectious diseases. Hard on the heels of the completion of about a dozen whole genome sequences of bacteria are plans to sequence the chromosomes of parasites such as *Plasmodium falciparum*. Increasingly, investigators will have available a catalogue of gene sequences that include every virulence factor, every potential drug target and every vaccine candidate for a particular pathogenic microbe. In addition, the explosion of sequence data will have major implications for population biologists, ecologists and evolutionary biologists.

The availability of the whole genome sequence of *Haemophilus influenzae* strain Rd (1.83 megabase pairs) has facilitated significant progress in characterising the biosynthetic pathway of its lipopolysaccharide (LPS). LPS is a critical structural and functional component of the cell envelope, a major virulence factor that is involved in every stage of the pathogenesis of serious *H. influenzae* infections such as meningitis (inflammation of the linings of the brain). By searching the *H. influenzae* genomic data base with sequences of known LPS biosynthetic genes from other organisms, we identified and then cloned 25 candidate LPS genes. This has allowed the construction of mutant strains and analysis of the LPS by reactivity with monoclonal antibodies and PAGE fractionation patterns. Electrospray mass spectrometry comparative analysis has confirmed a potential role in the LPS biosynthesis for the majority of these candidate genes. Studies in the infant rat have allowed us to investigate the role of LPS in pathogenicity and to estimate the minimal structure required for intra vascular dissemination. This is one of the first studies to demonstrate the power of whole genome sequencing to extend biological knowledge of a pathogenic microbe (*in silico* to *in vivo*). The speed and ease of detection of genes is significantly greater than that by classical molecular genetic analysis and, in particular, allows the identification of genes found even under circumstances of weak amino acid homology.

INTRODUCTION

Sequencing of DNA has brought about a revolution in biology by making available the immense fund of historical information contained in the genomes of cells. Unicellular organisms appeared some two million years before the first primitive algae and over a billion years before the first animals and higher plants. Thus, a sensible and obvious starting point in the sequencing of entire genomes might be to tackle the prokaryotes, the smallest chromosomes that contain all the information required for the free-living state. But, despite the enormous commitment to the human genome sequencing project throughout the 80s and early 90s, major funding agencies had been reluctant to support the sequencing of bacterial genomes. However, in 1995, the status of sequencing bacterial genomes changed dramatically. This milestone was achieved by a team led by Craig Venter and Robert Fleischmann at The Institute for Genome Research (TIGR) in Gaithersburg, Maryland, USA. The genome sequenced was that of *Haemophilus influenzae*, a major cause of invasive childhood infections including meningitis which was once mistakenly thought to cause epidemics of serious respiratory infections, now known to be caused by the influenza virus. The modest size (nearly 2 million base pairs) and high AT-nucleotide content of *H. influenzae* offered the perfect challenge in a bid to sequence the first complete genome. The project used the large-scale sequencing ability of TIGR. In a remarkable feat of expertise and team-work, the TIGR group assembled the sequence of 1.83 million nucleotides of circular double-stranded DNA in less than a year. A crucial factor in the success of the project was the application of a random sequencing approach and computational methods developed by TIGR for large-scale sequencing.

WHOLE GENOME SEQUENCING OF SMALL GENOMES: A REVOLUTIONARY TOOL IN MICROBIOLOGY

A crucial aspect of the approach taken to sequence the *H. influenzae* genome was the methodology. For most genome projects, the current strategy requires the construction of a physical map of overlapping cloned fragments and then these are sequenced. Venter, Smith and Hood (1996) argued that this approach to genomic sequencing was neither ideal nor necessary. Their alternative strategy involved sequencing a library of randomly sheared fragments (shot-gun approach), combined with computer informatics to assemble these fragments into a complete genome. This approach had been taken by Sanger (Sanger *et al.*, 1977; Sanger *et al.*, 1982) to complete the sequence of phage lambda (48,000 bp in size), but the use of this strategy to tackle a genome of almost 2 million nucleotide pairs (2Mb) was a bold step that depended on substantial expertise in bioinformatics. Despite initial scepticism, the assembled, annotated genome of 1,830,137 bp of *H. influenzae*, strain Rd, was completed in one year, an extraordinary and brilliant achievement (Fleischmann *et al.*, 1995). The success of the shot-gun approach depends critically upon ensuring that the genomic fragments to be sequenced are random, since any deviations from randomness are reflected in a proportionate increase in redundancy of the sequenced clones. Applying Poisson principles to a random set of sequenced clones, when a number of nucleotides equivalent to that of the genome have been sequenced, the probability is that about 37% of the genome will not be represented. After completing five times the number of nucleotides in the genome, less than 1% of the genome will remain unsequenced, but this still amounts to a considerable number of physical and sequence gaps. TIGR adopted a variety of ingenious approaches to the assembly process and this is where their previous experience in using and adapting sophisticated computer programming and informatics to facilitate assembly was crucial.

UTILITY OF WHOLE GENOME SEQUENCES

The sequencing of whole genomes affords a powerful and relatively economic way to make available a comprehensive bank of information on a wide range of organisms that will have many basic biological, medical, veterinary and industrial applications. For pathogenic bacteria, this information might be considered under several categories, a highly selective summary of which is considered as follows:

Linear sequence

Annotation of the *H. influenzae* genome by TIGR indicated 1,743 open reading frames (putative genes). Since the original publication in *Science* (Fleischmann *et al.*, 1995), many revisions to the sequence have been made such as those suggested by Brenner (Brenner *et al.*, 1995) and Tatusov (Tatusov *et al.*, 1996) In fairness to Fleischmann *et al.*, it is untenable to suppose that a completely accurate sequence for such a large molecule could be presented. Nonetheless, Fleischmann *et al.* estimated that the error frequency in their sequencing was between 1/5,000 and 1/10,000, i.e. less than 0.03%. It should also be noted that Fleischmann *et al.* took several novel steps to ensure that their sequences were well curated and updated by establishing a World Wide Web site as a central clearing house. This system is sensible and has been adopted for other projects.

Only one half to two thirds of the more than 1700 genes could be reasonably assigned functions. Many investigators have pointed out the problems associated with incautious conclusions drawn by those who assume functions from data base searches. Even when such predictions are undertaken critically, subsequent experiments may produce results which surprise or cause previous conclusions to be radically revised or abandoned. Nonetheless, many assignments could be made with confidence and included, for example, the various enzymes for intermediary metabolism, biosynthesis of small molecules, macromolecules, cell structure, cellular processes and other functions. These putative genes and their functions, depicted in helpful colour coded maps, provided an immediately accessible picture of the amount of DNA involved in different functions in *H. influenzae*, for example, 10% to energy metabolism, 17% to transcription and translation, 12% to transport and 8% to cell envelope proteins. These were the kind of assignments anticipated several years previously by Monica Riley (Riley, 1993). Whole genome sequences are revealing many new genes whose functions are not evident from inspection or use of homology searches. These novel genes represent an exciting challenge and an example of the richness of the "treasure trove". Thus, for an investment of about $1 million, a genome sequence identifies several hundreds of gene sequences among which are to be found those encoding every virulence factor, every target for therapy and every molecule that is a potential vaccine or target for host immune clearance mechanisms. Nonetheless, the systematic, inevitably selective, investigation of biological function represents a massive task.

In contrast to the genomes of higher (eukaryotic) organisms where up to about 95% of sequences are non-coding, it was correctly anticipated that a high proportion of the genome sequence of *H. influenzae* would code for proteins. But, non-coding DNA in *H. influenzae* was particularly interesting. Copies of motifs comprising 23 nucleotides are involved in DNA transformation, a mechanism whereby *H. influenzae* is able to recognise its own DNA. These nucleic acids are released following spontaneous lysis of sibling or closely related bacterial cells, and then incorporated into the genome of recipient cells by homologous recombination (Smith *et al.*, 1995). Another finding was the presence of multiple tandem tetrameric repeats within open reading frames of known or putative virulence factors (Hood *et al.*, 1996a). These repeated tetranucleotides, through polymerase slippage, have been shown to mediate antigenic variation by causing frame shifts within genes, for example, those involved in the biosynthesis of lipopolysaccharide and are considered in more detail later in this review.

Gene expression

Sequence information from a genome can be used to characterise not merely the activities of individual genes but opens the door to a more comprehensive investigation and understanding of the integration and coordination of genotype and phenotype under different conditions. For example, altered gene expression associated with bacterial entry into host cells as compared to their extracellular residence. This genotypic analysis might be undertaken using a library of genes representative of the whole genome, whose complete sequence is known, arrayed on a solid phase and probed with labelled RNA from bacteria isolated and grown under different conditions. To complement this genotypic approach, there have been many advances in techniques, such as the combination of two-dimensional gel electrophoresis and mass spectrometry, with which to take global approaches to the analysis of the gene products and their post-translational modifications.

Population biology

Conventional taxonomy in microbiology has relied upon phenotypic characterisation which, although useful, may obscure or frankly distort the appreciation of genetic relationships and population structure. Multi-locus enzyme electrophoresis was recruited to investigate population structures of bacteria, focussing particularly on pathogenic species (Selander and Musser, 1990). This indexes genic variation through comparison of polymorphisms in a sample of metabolic (housekeeping) enzymes. It is a powerful tool for studying population structure. It allows large numbers of organisms to be characterised in a relatively short period and is amenable to statistical analyses that allow strong inferences to be drawn concerning the extent of horizontal gene transfer and inter genomic recombination among strains of a species. With the advent of automated sequencing, a two pronged strategy can be applied in which the DNA sequence of particular virulence genes can be compared in strains selected from a population framework defined by multilocus enzyme electrophoresis. These phylogenetic frameworks have in a few instances been supplemented by more detailed analyses of the inter-strain variation in nucleotide sequence data of selected genes. Whole genome sequences provide yet another level of resolution in that virulence is multi factorial and must be considered in the context of the eclectic contribution of many genes including the molecules that determine tropism, *in vivo* replication and cytotoxic factors. New methods are clearly needed for these types of analyses. For example, high density oligonucleotide arrays for large scale analysis of genome data may provide the technical means to achieve these aims (Goffeau, 1997). The perspective offered by the availability of whole genome sequences will, *per se*, offer a more informed vantage point from which to study populations. Some major gaps in our knowledge include the extent of the conservation of coding and non-coding nucleotides, the extent of rearrangements and intra genomic mobility and the extent and rates of recombination. These data have a very practical application in determining the variation in molecules that are targets for drugs and vaccines in the control of microbial infections.

Comparative genomics

One of the most exciting avenues of research opened up by whole genome sequences is that of comparative biology. It is now possible to index not only the variations within and between species, but also to derive an estimate of the minimal set of genes required for the essential functions of a cell and to trace back the gene composition of ancestral organisms. A practical application of this will be the ability to undertake rational attenuation of bacteria and their use as live vaccines or vectors for foreign antigens.

Pathogenicity

The impact of whole genome sequencing of bacteria is particularly exciting and relevant to the field of infectious diseases. Major pathogens, such as *Mycobacterium tuberculosis* and *Streptococcus pneumoniae*, are the cause of much morbidity and mortality. New approaches to therapy through novel drugs or the development of vaccines may be discovered or facilitated through the information contained in whole genome sequences. In particular, the high proportion of previously undiscovered genes (for some of which there are as yet no clues as to their function) constitutes an exciting challenge and potential opportunity for novel pharmaceuticals or biologicals.

USE OF WHOLE GENOME SEQUENCE TO INVESTIGATE THE LIPOPOLYSACCHARIDE OF *H. INFLUENZAE*

The availability of the complete genome sequence of *H. influenzae* affords an example of how the information can be used to go from "*in silico* to *in vivo*". The molecule that we have studied is the lipopolysaccharide (LPS) of *H. influenzae*, a critical structural and functional component of the bacterial cell envelope (Moxon and Maskell, 1992). LPS is known to be a key factor in the pathogenic potential of *H. influenzae*. LPS modulates the surface interactions of *H. influenzae* with host cells (e.g. epithelial cells of the respiratory tract) and is important in the potential of *H. influenzae* to translocate across the tissue barriers of the respiratory tract. In addition, LPS plays a key role in the organism's ability to withstand clearance from the blood-stream. Lipid A (endotoxin) of LPS is a potent mediator of inflammation, e.g. in pneumonia or meningitis. LPS is a complex glycolipid and its biosynthesis requires many genes (more than 40) to make, transport and assemble the sugars, fatty acids and other substituents (e.g. phosphoethanolamine) of the mature, surface located molecule. Prior to the availability of the complete genome of *H. influenzae* (strain Rd), only a few genes for LPS had been identified (Hood, *et al.*, 1996b). Using previously published sequences for lipopolysaccharide genes (or their translated proteins) available in the general data bases, these "probes" (either DNA or deduced amino-acid sequence) were used to search the *H. influenzae* genome for homologous sequences. If matches ("hits") to open reading frames were found, these candidate genes were cloned and characterised. In particular, site directed mutations were made by inserting a cassette of DNA encoding kanamycin resistance within the open reading frame to disrupt the gene. Comparison of the mutant and wild-type gene in the isogenic strain of *H. influenzae* provided a powerful strategy for assaying changes in phenotype brought about by mutations of the gene under investigation. We used gel electrophoresis and silver staining to assess possible differences in LPS migration characterisation, reactivity with a panel of monoclonal antibodies and fine structural analysis of extracted LPS using electrospray mass-spectrometry (Masoud, *et al.*, 1997; Fleischmann, *et al.*, 1995). Studies in an animal model of *H. influenzae* infection were then used to investigate the relative effects of structural differences in LPS on virulence. These indicated the role of chain extensions (outer core saccharides) to a basal structure (inner core) in mediating intra vascular survival of *H. influenzae* organisms during bacteremia (figure 1) (Hood, *et al.*, 1996b). The extent of conservation of inner core structures among all strains of *H. influenzae* can now be investigated to provide a rational approach to the identification of structures that may be targets for host immunity. The idea is to determine the candidacy of these inner core structures in broad range vaccine development.

In summary, the whole genome sequence of *H. influenzae*, strain Rd, has provided information that has facilitated progress on the genetics, structure and function of LPS. Once available, the genomic approach offered substantial advantages over classical genetics in both the rapidity and efficiency with which these investigations could proceed.

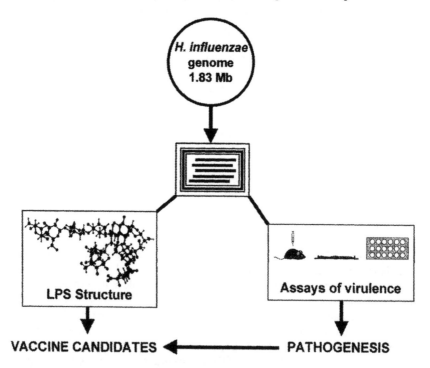

FIGURE 1. Utility of whole genome sequencing of bacterial pathogenesis

REFERENCES

Brenner, S.E., Hubbard, T., Murzin, A. and Chothia, C. (1995) Gene duplication in *H. influenzae*. *Nature* **378**, 140.

Fleischmann, R.D., Adams, M.D., White, O., Clayton, R.A., Kirkness, E.F., Kerlavage, A.R., Bult, C.J., Tomb, J.-F., Dougherty, B.A., Merrick, J.M., *et al.* (1995) Whole-genome random sequencing and assembly of *Haemophilus influenzae* Rd. *Science* **269**, 496-512.

Goffeau, A. (1997) Molecular fish on chips. *Nature (News and Views)* **385**, 202-203.

Hood, D.W., Deadman, M.E., Allen, T., Martin, A., Brisson, J.R., Fleischmann, R., Venter, J.C., Richards, J.C. and Moxon, E.R. (1996b) Use of the complete genome sequence information of Haemophilus influenzae strain Rd to investigate lipopolysaccharide biosynthesis. *Molecular Microbiology* **22**, 951-965.

Hood, D.W., Deadman, M.E., Jennings, M.P., Bisercic, M., Fleishchmann, R.D., Venter, J.C. and Moxon, E.R. (1996a) DNA repeats identify novel virulence genes in *Haemophilus influenzae*. *Proceedings of the National Academy of Sciences, USA* **93**, 11121-11125.

Masoud H., Moxon ,E.R., Martin, A., Krajcarski, D. and Richards, J.C. (1997) Structure of the variable and conserved lipopolysaccharide oligosaccharide epitopes expressed by *Haemophilus influenzae* serotype b strain Eagan. *Biochemistry* **36**, 2091-2103.

Moxon, E.R. and Maskell, D. (1992) *Haemophilus influenzae* lipopolysaccharide: the biochemistry and biology of a virulence factor. In: *Molecular biology of bacterial infection: current status and future perspectives,* C. Hormaeche, C.W. Penn, C.J. Smyth (Eds.), *Society for General Microbiology Symposium,* **49**, 75-96.

Riley, M. (1993) Functions of the gene products of *Escherichia coli*. *Microbiological Reviews* **57**, 862-952.

Sanger, F., Coulson, A.R., Hong, G.F., Hill, D.F. and Petersen, B., (1982) Nucleotide sequence of bacteriophage lambda DNA. *Journal of Molecular Biology,* **162**, 729-773.

Sanger, F., Air, G.M., Barrell, B.G., Brown, N.L., Coulson, A.R., Fiddes, C.A., Hutchison, C.A., Slocombe, P.M. and Smith, M. (1977) Nucleotide sequencing of bacteriophage phi X174 DNA. *Nature,* **246**, 687-695.

Selander, R.K. and Musser, J.M. (1990) The population genetics of bacterial pathogenesis. In: *Molecular Basis of Bacterial Pathogenesis*, V.L. Clark, B.H. Iglewski (Eds.), Orlando, Florida: Academic Press, Inc.. pp11-36.

Smith, H.O., Tomb, J.F., Dougherty, B.A., Fleischmann, R.D. and Venter, J.C. (1995) Frequency and distribution of DNA uptake sequence sites in the Haemophilus infleunzae Rd genome. *Science,* **269**, 1-3.

Tatusov, R.L., Mushegian, A.R., Bork, P., Brown, N.P., Hayes, W.S., Borodovsky, M., Rudd, K.E. and Koonin, E.V. (1996) Metabolism and evolution of *Haemophilus influenzae* deduced from whole-genome comparison with *Escherichia coli*. *Current Biology* **6**, 279-291.

Venter, J.C., Smith, H.O. and Hood, D. (1996) A new strategy for genome sequencing. *Nature* **381**, 364-366.

PHARMACOGENETICS: PRINCIPLES AND APPLICATIONS

GLENN A. MILLER, Ph.D.

Molecular Profiling Laboratory, Genzyme Genetics, P.O. Box 9322, Framingham, MA 01701-9322, USA   email:gmiller@genzyme.com

ABSTRACT

Recent advances in the study of genetic-based responses to therapeutic compounds have increased interest in the use of genotypic information to segment patient populations. The role that pharmacogenetics will play in future drug development will depend upon further advances from the Human Genome Project and functional genomics. The future of drug development will require detailed knowledge of genetic factors important to drug targeting, response and effectiveness. The following review will present some of the current challenges and opportunities that pharmacogenetics presents to the development of the next generation of therapeutic compounds.

INTRODUCTION

There are no completely innocuous therapeutic compounds, only those compounds whose side effects are outweighed by their positive therapeutic value. The greatest potential for pharmacogenetics involves the use of genotype/phenotype information, obtained from affected populations, to design therapeutic modalities effective on the majority of individuals with minimal side effects. Pharmacogenetics offers the possibility of designing and prescribing therapy based on the likely response of an individual as determined by their genotype. The current paradigm for pharmacogenetics is to use polymorphisms affecting the expression of genes of known function to segment a population into various categories, as determined by the efficiency at which a compound is metabolized. The future of pharmacogenetics will use the genes and pathways defined by the Human Genome Project as the raw material of compound design, creating reagents with high therapeutic indices of identifiable market value in well-characterized populations. The route to this new paradigm of therapeutic design will require the use of the full range of molecular genetic technology. There will be no single technique or assay which will provide all of the information required for the segmentation of patient populations. The identification of relevant genes and polymorphisms will require the concerted use of a variety of techniques able to easily identify known polymorphisms, as well as techniques able to scan genes for novel polymorphisms of functional importance. The biotechnology industry is well positioned to assist in the future that pharmacogenetics presents to the pharmaceutical industry. This new paradigm of pharmaceutical development will only become reality, however, if the information that the Human Genome Project and the biotechnology industry provides is relevant to the creation of compounds of greater therapeutic efficacy and broad market acceptance.

This review will focus on the current uses of pharmacogenetics and how current technology can facilitate the advance of this new paradigm of pharmaceutical development.

PHARMACOGENETICS TODAY

The greatest use of pharmacogenetics in current therapeutic development is in the segmentation of a population by their relative efficiency of compound metabolism. There are a small number of genes that encode enzymes having been identified as important to metabolic pathways for the majority of compounds in use today. Table 1 lists the most commonly studied of these enzymes TABLE 1. Enzymes important in metabolic pathways of common pharmaceutical compounds.

Enzymes involved in Drug Metabolism
CYP1A2
CYP2A6
CYP2B
CYP2C19
CYP2D6
CYP2E1
CYP3A
NADPH-quinone
Oxidoreductase
Glutathione S-transferase
N-acetyltransferase
UDP-glucuronosyltransferase
Sulfotransferases

Historically, pharmacogenetic studies have involved the phenotypic evaluation of a population of individuals. Such information could, until recently, only be derived by the administration of a variety of drugs known to be metabolized by a certain pathway or enzyme. The information derived from such studies has resulted in the creation of three categories of metabolizers: poor, extensive and ultra-extensive (PM, EM and UEM, respectively). The ethical implications, however, of administering a bioactive compound to an individual in the absence of a clinical need has limited many of these studies to small populations. The cloning of the CYP2D6 gene in 1989 by Kimura, et al. allowed the genotyping of large numbers of individuals and the subsequent correlation of genotype with phenotype. This has allowed a number of studies to be completed without the ethical constraints involved with clinically-unnecessary drug administration.

CYP2D6 accounts for the metabolism of as much as 25% of all commonly prescribed drugs. As a result, this enzyme, and its alleles, has become one of the most intensely studied metabolic enzymes in pharmaceutical development.

Cytochrome P450 2D6

A member of the cytochrome P450 family of monooxygenases, CYP2D6 is a good example of how metabolic enzymes can be used to segment populations. A range of metabolic phenotypes have been identified and correlated with particular CYP2D6 genotypes. A total of 17 alleles have been identified and characterized with respect to enzyme activity. A difficulty in the design of assays for CYP2D6 is the presence of two pseudogenes in the immediate vicinity. This results in several of the 2D6 alleles being formed through gene conversion or the formation of a hybrid between the gene and a pseudogene. Sachse, *et al.* (1997) have designed a series of nested PCR reactions able to characterize all of the known alleles of CYP2D6. Of clinical importance is the finding by Sachse, *et al.* (1997) of a limited number of alleles as characterizing the poor metabolizing phenotype. Therefore, assays to segment a population into poor versus extensive metabolizers should be able to be constructed in a cost-efficient and high throughput manner.

One method for rapidly analyzing the majority of alleles in the CYP2D6 gene is an assay developed by Shuber, *et al.* (1997), named MASDA for Multiplex Allele Specific Diagnostic Assay. This assay permits the simultaneous examination of hundreds of alleles in hundreds of patient samples. Making use of Allele Specific Oligonucleotides (ASOs) and tetramethyl ammonium chloride (TMAC) hybridization chemistry, MASDA permits the pooling of ASOs under identical hybridization conditions. In a forward dot blot format, this assay allows hundreds of patient samples to be assayed simultaneously. The patient sample acts to hybrid select the appropriate allele from the pool of ASOs in solution. The spot containing the patient sample is then physically removed from the membrane and the bound ASO(s) eluted and identified by any one of a number of techniques, including sequencing.

Cytochrome P450 2D6 is but one of a number of genes potentially useful in the stratification of patient populations. As the Human Genome Project advances, a growing number of genes and polymorphisms will be determined to be of relevance to a variety of pharmaceutical programs. The discipline of pharmacogenetics will become a critical component of drug development and administration.

PHARMACOGENETICS IN THE NEAR TERM

Pharmacogenetics currently involves the use of a limited set of reasonably well-understood genes and pathways. The types of studies currently being performed are often merely supplements to other biochemical tests routinely used in clinical trials. The next step for pharmacogenetics will use genetics as a first rank tool in clinical trial design and evaluation. There are a number of areas where pharmacogenetics can play a role in drug development in the near future.

Inclusion criteria

The spectrum of drug efficacy and tolerance across a population is often the result of various genetically-determined metabolic rates within a population. The role that CYP2D6 plays in common drug metabolism is a clear example of the role that genetics can play in drug efficacy. A role for pharmacogenetics in the near future will likely be the inclusion or exclusion of individuals from a clinical trial based on previously determined genotype information. Knowledge of how a particular genotype can alter or prevent the side effects seen in a trial population could result in the trial design being altered to specifically exclude or include the genotype under study. The use of this information could, therefore, significantly alter the accepted practices for effectiveness study design. In recognition of the advancements being made in clinical trial design the U.S. Food and Drug Administration(FDA) has circulated a draft guidance document for comment(Docket Number 97D-0100). This document is meant to open preliminary discussions concerning the amount and rigor of clinical evidence in support of effectiveness claims. While not meant for implementation, the Guidance provides an interesting perspective into the possible future for pharmacogenetics with respect to effectiveness demonstration

Effectiveness demonstration

The Federal Food, Drug and Cosmetic Act of 1938 was amended in 1962 to include an effectiveness requirement for drug approval. This amendment required drug manufacturers to demonstrate that not only was their drug safe, but effective as well. Since the adoption of this amendment the methods and practices of drug development have improved significantly. These improvements are likely to alter the type and quantity of data required to prove effectiveness. It is likely in the future that studies will be altered to define more narrowly clinical criteria or identify a distinct sub-population. As one defines a clinical trial's focus more narrowly it is conceivable that other equally well-defined studies will become available to use in supporting a particular compounds effectiveness application.

The FDA has traditionally required that "two adequate, well-controlled studies, each convincing on its own" be required to demonstrate effectiveness (FDA, Docket Number 97D-0100). Effectiveness information already derived from a given patient population may, in the opinion of this draft document, be used as a foundation for another study. Thus the data from a single study would then, in conjunction with the data already known, be sufficient to support effectiveness in another subset of that population. The ability of pharmacogenetics to provide detailed population-based information would then be the scientific underpinnings for the use of data from other accepted clinical studies in the support of drug effectiveness.

Expanding the indications for therapeutic compounds is another area where pharmacogenetics could be used to great effect. The FDA guidance document states that "information about drug pharmacology, such as the extent of drug uptake or distribution in biochemical or metabolic processes or known information about receptor binding in a specific disorder, might indicate it would be appropriate to extend the results of a single diagnostic study to a related disorder". As the molecular mechanisms of drug action and interaction are elucidated, the amount of information relevant to the extension of results of one study to that of another study of a related disorder will become more frequent.

The use of pharmacogenetic data in support of effectiveness studies, and the expansion of indications for therapeutic compounds, are the most likely early-impact areas for pharmacogenetics. The true value of pharmacogenetics might be more appropriately termed pharmacogenomics. It is in the information that the Human Genome Project and the vast array of cloning and expression genetics efforts will provide that pharmacogenetics will truly become an operating tool of pharmaceutical development.

THE FUTURE OF PHARMACOGENETICS

Pipeline requirements and new paradigms

Jurgen Drews of Hoffmann-LaRoche Pharmaceuticals predicts that the Human Genome Project may produce 3,000 to 10,000 novel drug targets of interest over the next six years. The reliance of the pharmaceutical industry on these new targets is apparent when one considers the requirement by the industry for at least 42 new chemical entities (NCEs) per year in order to maintain a 10% growth rate. The Human Genome Project will not be able to provide all of the targets necessary to yield the full allotment of NCEs required by the pharmaceutical industry. In this light it is apparent that a new paradigm of drug development must be created. This new paradigm will make use of a variety of resources to fuel the NCE pipeline.

The future of drug development will make use of the vast array of information sources now being developed, including the Human Genome Project, functional genomics, population-based pharmacogenetics, rational drug design and high throughput screening, using targets specifically designed to mimic the targets of increasingly complex disease states.

One scenario for the future of drug development is depicted in figure 1.

While from a production sequencing point of view the Human Genome Project may be completed over the next decade, the effort required to extract usable data from this data repository will continue well into the next century. The discipline of functional genomics holds great promise in the arena of target identification and lead discovery. The major role for functional genomics, however, is in the understanding of molecular pathogenesis based on the elucidation of both normal and disease-state function of a given gene or set of genes. As more and more complex disease states are identified as legitimate therapeutic targets, a more complete understanding of the genetic pathways and interactions of various proteins and genes will be required.

Identification of disease mechanism and/or genetic basis of disease
/
Elucidation of targets for therapeutic intervention
/
Rational drug design specific for target
/
Stratification of patient population into segments likely to respond to therapy
/
Narrowly designed clinical trials based on population segmentation
/
Effectiveness demonstrated in multi-center trials and from information derived from other studies definitively linked through pharmacogenetic information
/
Drug launch

FIGURE 1. A proposed schematic drug development pathway given information derived from the Human Genome Project, functional genomics and pharmacogenetics.

Gene expression studies.   The appetite for drug targets and lead discovery by the pharmaceutical industry, as a group, will be fed, in part, by the advancing technology in differential gene expression.   A number of companies and academic laboratories are involved in the development of technologies to measure the level of gene expression in either a group of known genes or by sampling all of the expressed genes in a tissue sample. Generally speaking, the study of gene expression falls into two major categories.

One category involves those technologies that measure gene expression in a group of known expressed genes. This technology is most commonly viewed as an array of gene segments on a solid support, the DNA chips and other arrays. The number of different genes whose expression can be profiled in this manner is very large but finite. Some information must be known concerning the genes placed in the arrays in order to synthesize the probes from which the arrays are constructed. This bias towards known genes, no matter how well characterized, limits the ability of these arrays to identify correlations in gene expression among unknown and rarely-expressed genes. An additional difficulty with the DNA chip arrays is their inflexibility with regard to the addition of newly-discovered genes of importance to a particular disease state. The development time and costs for each chip mitigate against the regular and fluid alteration of the sequences arrayed upon them.

The second category of expression technologies in use today is the analysis of all expressed genes in a tissue with the subsequent comparison between tissue types or disease states. A technique pioneered by Velculescu, et al. termed Serial Analysis of Gene Expression (SAGE) permits the analysis of large numbers of expressed genes in a serial manner. The technique is based on two principles: 1) A short nucleotide sequence (9-10 base pairs in this instance) is sufficient to unambiguously identify a transcript and 2) by concatenating these short sequence tags the analysis of expressed genes can take place in an efficient, serial manner by sequencing. Through the use of SAGE it is possible to analyze all of the expressed genes in a given tissue without prior knowledge of any of the gene sequence, structure or function. The depth of the tag library that is constructed determines the statistical likelihood of identifying a tag of a rarely expressed gene. This is in contrast to DNA arrays which cannot detect previously uncharacterized or unknown genes, regardless of their rarity of expression.

Gene expression studies are important to lead discovery and drug development in addition to the raw sequence data from the Human Genome Project. A significant example of the additive power of expression studies to full sequence data is the work of Velculescu, *et al.*. This study examined all of the expressed sequences in *S. cerevisiae* using the SAGE technology. There were a total of 4,665 genes identified by SAGE. Upon analysis of the databases then available it was observed that 1,981 of the genes had previously identified functions. The remaining 2,684 genes were unknown at the time of analysis. In addition, the total number of genes identified was larger than would have been predicted using the genomic sequence alone. It is, therefore, apparent that other studies like expression pattern and functional studies must accompany the raw sequence data generated through the Human Genome Project.

SUMMARY

The future of drug development will require an increasing number of targets and lead compounds. The current climate of managed care and cost reduction pressures will demand that the selection process for lead compounds, as well as drug targets, have a much higher rate of success than is currently the norm. The traditional methods of drug discovery, consisting of large-scale screening of extensive libraries of chemical compounds against targets of often questionable relevance to the therapeutic target, must adapt to accommodate this changing environment. The level of information provided by such resources as the Human Genome Project, functional genomics and pharmacogenetics will permit a wholesale overhaul of the drug design process. It is unlikely, however, that this information will be used to design therapeutic compounds for patient populations of decreasing size. The concept that a market segment will be increasingly subdivided to avoid untoward side-effects is a limited one. It is more likely that those pharmaceutical companies that make extensive use of genomics and genotypic information will use that information to design or select compounds with the least overall negative value for all potential patients. Given the choice between a lead compound that will well serve a small sub-population and a compound that is likely to serve a broad population, the choice will be to select the broadest market possible. Pharmacogenetics will then be brought to bear to select the clinical trial population that most appropriately fits the potential patient population while excluding those sub-populations that may not respond as desired. Inherent in this scenario is the assumption of a great deal of prior knowledge regarding the disease process, the metabolic pathways with which the therapeutic interacts and the population genetics of those to be treated. This level of knowledge can only be achieved via a close collaboration between major pharmaceutical companies and their biotechnology and academic partners.

The identification of lead compounds and selection of viable candidates through the use of genomics technology of various types is of great significance to the long range future of drug development. The role that pharmacogenetics can play in drug development is not confined to this future a decade or more away. Through the segmentation of patient populations and the design of more efficient clinical trials, pharmacogenetics can have an immediate impact on the approval process for drugs currently well down the development path.

The role that pharmacogenetics will play in drug development will be driven by technology only to the extent that the information provided serves the goals of the project. The cost that pharmacogenetic studies will add to clinical trials and drug development must be counterbalanced by the value that the information contributes to the project. The current economic environment in the health care industry will not support unlimited studies of population genotypes in order to design therapeutic modalities that serve increasingly small patient populations at increased cost. The greatest use of pharmacogenetics will be in aiding the efficient design of low cost therapeutics serving the broadest possible population.

REFERENCES

Benet, L., Kroetz, D. L. and Sheiner, L. B. (1996). Pharmacokinetics. *The Pharmacologic Basis of Therapeutics*. Eds. J. Hardman, Goodman, Gilman A., Limbird, L.E. New York, McGraw-Hill: 3-27.

Daly, A., Brockmoller, J., Broly, F., Eichelbaum, M., Evans, W. E., Gonzalez, F. J., Huang, J.D., Idle, J. R., Ingelman-Sundberg, M., Ishizaki, T., Jacqz-Aigrain, E., Meyer, U. A., Nebert, D. W., Steen, V. M., Wolf, C. R. and Zanger, U. M. (1996) Nomenclature for human CYP2D6 alleles. *Pharmacogenetics* **6** 193-201.

Drews, J. (1996) Genomic sciences and the medicine of tomorrow. *Nature Biotechnology* **14** 1516-1518.

Kimura, S., Umeno, M., Skoda, R. C., Meyer, U. A. and Gonzalez, F. J. (1989) The human debrisoquine 4-hydroxylase (CYP2D) sequence and identification of the polymorphic CYP2D6 gene, a related gene and a pseudogene. *American Journal of Human Genetics* **41** 889-904.

Linder, M., Prough, R. A. and Valdes, R. (1997) Pharmacogenetics: a laboratory tool for optimizing therapeutic efficiency. *Clinical Chemistry* **43**(2) 254-266.

Sachse, C., Brockmoller, J., Bauer, S. and Roots, I. (1997) Cytochrome P450 2D6 variants in a caucasian population: Allele frequencies and phenotypic consequences. *American Journal of Human Genetics* **60** 284-295.

Shuber, A., Michalowsky, L. A., Nass, G. S., Skoletsky, J., Hire, L. M., Kotsopoulos, S. K., Phipps, M. F., Barbeiro, D. M. and Klinger, K. W. (1997) High throughput parallel analysis of hundreds of patient samples for more than 100 mutations in multiple disease genes. *Human Molecular Genetics* **6**(3) 337-347.

Velculescu, V., Zhang, L., Vogelstein, B. and Kinzler, K. W. (1995) Serial analysis of gene expression. *Science* **270** 484-487.

Velculescu, V., Zhang, L., Zhou, W., Vogelstein, J., Basrai, M. A., Bassett, D. E., Heiter, P., Vogelstein, B. and Kinzler, K. W (1997) Characterization of the yeast transcriptome. *Cell* **88** 243-251.

APPLICATIONS IN AGRICULTURE

IMPACT OF GENOMICS ON IMPROVING CROP PROTECTION TECHNOLOGY

J. RIESMEIER

PlantTec Biotechnology Limited Research and Development, Hermannswerder 14, 14473 Potsdam, Germany

L. WILLMITZER, T. ALTMANN

Max-Planck-Institute of Molecular Plant Physiology, Karl-Liebknechtstr. 25 (Haus 20), 14476 Golm, Germany

ABSTRACT

In the next two or three decades, food and feed production will have to be raised dramatically due to a world-wide increase of the human population and changes of consumer behaviour. On the other hand, the total expanse of land under cultivation will stay constant or may even decline. Farmers, breeders and agrochemical industry together have to face this challenge. It is now necessary to evaluate and use all available traditional and new techniques to increase crop yield on an ecologically and economically favourable basis.

INTRODUCTION

In the last fifty years modern breeding methods and the development and use of effective crop protection chemicals have led to a substantial increase in crop yield. To keep this relative increase at the same rate as in the past, an ever-increasing effort with respect to both time and money is necessary.

The agrochemical world market has remained constant over the last several years at approximately US$ 25 billion per year. 1996 saw the arrival of a new agricultural market: genetically modified plants. In North America, 1.9 million hectares of transgenic corn, canola, soybean, cotton and potatoes were planted in 1996. This year a dramatic 12.6 million hectares will be on the fields, about 15 % of the total acreage covered by these crops. Most first-generation transgenic plants exhibit herbicide-resistance or insect-tolerance traits. Herbicide resistant crops allow a return on investment by increasing the market-share of the respective herbicide. The main players in this field are AgrEvo (glufosinate-resistant crops) and Monsanto (glyphosate-resistant crops). On the other hand, insect-tolerant plants which express a Bt toxin are sold without a companion product by seed companies. As the insect tolerance to these crops reduces the input of insecticides and thereby the farmers costs, an increased price for the seeds is acceptable to the customer. A simple calculation using a surcharge of US$ 5 per hectare of transgenic crops amounts to a potentiall annual turn-over of more then US$ 1 billion in North America alone (250 million hectares total farmland).

The introduction of new traits like herbicide and insect resistance is relatively straight forward. In such cases, a heterologous gene encoding a herbicide detoxifying activity or the familiar Bt toxin is expressed. However, it will require a joint effort of (bio)chemists, biologists, geneticists and breeders to solve other more complex agronomic problems, such as resistance to pathogenic bacteria or fungi, drought, salinity or others. Genome analysis ("genomics") should to be considered as a powerful tool in identifying target genes and proteins with respect to resistance and yield.

PLANT GENOMICS

Genomics covers both the structural elucidation of the organisation of a genome and the assignment of the function(s) that are carried out by the full complement of genes and, to a rather minor extent, by those regions responsible for the maintenance of the integrity of the genome or those providing the flexibility required to allow changes to occur. The structural genome analysis involves cytogenetic characterisation, the generation of a genetic map, the establishment of a physical map and, as the highest level of resolution, the determination of the complete DNA sequence. While the efforts (time and costs) required to achieve the generation of a physical map and the elucidation of the genome sequence largely depend on the size (complexity) of the genome and thus can be calculated fairly well, the functional analysis covers a yet unforeseeable complexity as it principally aims at describing the role of every individual gene or its product in the course of the events that take place during the life cycle of a plant exposed to varying environmental conditions. It is the gain of the latter information (based on the structural genome data) that will provide an entirely new level in the understanding of the morphogenetic and metabolic processes occurring in plants. This information is not only valuable for the scientific community but also offers a tremendous potential for commercial applications. Realising these opportunities has led not only to a fast and ongoing development of new techniques for the structural genome (sequence) analysis, but also to the generation of novel resources designed to assign functions to the genes identified by virtue of their sequence.

Methods and techniques in plant genomics

Recent advances in the structural genome analysis, which have been achieved to a varying extent for different plant species, involve the introduction of molecular markers, the establishment of physical maps and the large scale generation of genomic and cDNA sequences. Molecular markers such as RFLPs (restriction fragment length polymorphisms; Botstein et al., 1980), RAPDs (random amplified polymorphic DNAs; Williams et al., 1990) CAPS (co-dominant cleaved amplified polymorphic sequences; Konieczny and Ausubel, 1993), microsatellites or SSLPs (simple sequence length polymorphisms, Hearne et al., 1992), and AFLPs (amplified fragment length polymorphism; Vos et al., 1995) plus derivatives thereof, provide efficient means for the determination of the genome composition, e.g. after crosses between different genotypes and provide the contact/reference points between the genetic map and the physical map. The availability of a relatively limited set of molecular markers has already allowed the advent of a new era of quantitative genetic analysis (Tanksley, 1993). Thus, so-called QTL (quantitative trait loci) can now be assigned to certain regions of the genome that contribute to important traits such as yield, stress tolerance (biotic and abiotic), heading date/flowering, fruit size and quality. Furthermore, these markers greatly facilitate and speed-up backcross programs performed to introgress certain favourable traits (e.g. disease resistance) into elite genotypes and thus advanced already to indispensable tools for "traditional" plant breeding. The availability of simple and reliable molecular marker systems in combination with sophisticated QTL analysis tools is expected to enhance future breeding programs, through the application of marker-assisted selection or even marker.driven selection. Another hallmark of the utility of molecular markers is the recent achievements in map-based cloning of genes which could only be defined phenotypically (the expression of a defined phenotype as the consequence of a genetic alteration). Besides the application of insertion mutagenesis (see below), this technology provided the only accession at the molecular level to genes responsible for traits as important as disease resistance (e.g. Martin *et al.*, 1993), flowering time (Putterill *et al.*, 1995), or plant growth regulator signalling (e.g. Giraudat *et al.*, 1992). The spectrum of plant species used for this work reflects the degree of genome characterisation achieved for various plant species (in combination with their accessibility to genetic transformation) with *Arabidopsis*, tomato and rice being most advanced with average marker densities of about 1 per cM. Hitherto, only very few genes have been isolated by this technique from plant species with less dense molecular marker maps, like sugar beet and/or barley. In both cases, the isolated genes confer pathogen resistance (Cai *et al.*, 1997; Büschges *et al.*, 1997). Increased maker densities combined with the development of substitution lines (near isogenic lines) are expected to provide the means to isolate even QTLs at the molecular level through map-based cloning (Tanksley *et al.*, 1995).

A high over-all marker density is of even greater importance for the generation of a physical map composed of ordered arrays of overlapping cloned DNA fragments (contigs) which represent the complete genome. While relatively short ranged contigs are being constructed in the course of map-based cloning approaches for a (still restricted) number of plant species, the generation of global physical maps covering the whole genome are currently being generated essentially for two model systems: *Arabidopsis* and rice. In both cases, large insert genomic libraries (YAC, BAC, P1) have been generated (Goodman *et al.*, 1995; Mozo *et al.*, 1996; Wang *et al.*, 1995; Zhang *et al.*, 1996) whose availability is another prerequisite for physical mapping work. These efforts already resulted in YAC physical maps for the *Arabidopsis* chromosomes 4 (Schmidt *et al.*, 1995) and 2 (Zachgo *et al.*, 1996). The establishment of YAC physical maps for the remaining 3 *Arabidopsis* chromosomes is in an advanced state with more than 80% of the *Arabidopsis* genome covered
(see http://cbil.humgen.upenn.edu/~atgc/ATGCUP.html,
http://genome-www.stanford.edu/ *Arabidopsis*/) and the generation of a BAC physical map is in progress (see http://probe.nalusda.gov:8000/otherdocs/pg/pg5/abstracts/p-5e-207.html). Likewise, more than 50% of the rice genome has been covered by YACs identified by mapped markers
(see http://probe.nalusda.gov:8000/otherdocs/pg/pg5/abstracts/w-rice-64.html). This information is not only of high value for future map-based cloning efforts but also serves, in the case of *Arabidopsis*, as a basis for the generation of high resolution BAC-contigs. These contigs as well as the information gained from BAC end-sequencing are used to coordinate the international *Arabidopsis* Genome Initiative (AGI; see http:genome-www.stanford.edu/Arabidopsis/AGI) which use existing BAC and P1

clones to sequence the entire *Arabidopsis* genome (Bevan *et al.*, 1997). With about 4.5 Mbp genomic sequence hitherto obtained and the addition of further 12 - 15 Mbp per year, the complete sequence of the low copy regions of the *Arabidopsis* genome (about 100 Mbp) will be completed by this joint effort by the end of 2003 at a cost of about $50-75 million. 1.8 Mbp of contiguous genomic sequence generated by a consortium of 18 EU labs indicated a high density of genes with one putative gene every 4.7 Mbp most of which have not been identified before (see http://probe.nalusda.gov:8000/otherdocs/pg/ pg5/abstracts/s25.html). Another global sequencing effort aimed at the identification of *Arabidopsis* genes has been initiated even earlier: the determination of end sequences of cDNAs picked from various cDNA libraries (Newman *et al.*, 1994; Delseny *et al.*, 1997) resulted in the accumulation of more than 30,000 ESTs (expressed sequence tags) that probably represent 30 - 40% of all *Arabidopsis* genes. This extensive resource already had a dramatic influence on plant molecular genetic analysis with a very large set of genes isolated and studied not only in *Arabidopsis* but also in various other plant species, only a minor fraction of which could have been isolated by traditional methods. ESTs are furthermore generated for rice (currently about 13,000 ESTs), maize (currently about 1,800 ESTs), oilseed rape (currently about 1400 ESTs) and a few other plant species (currently less than 1,000 ESTs each).

The overwhelming amount of sequence information which will be available within the near future sets an even more ambitious task: the functional characterisation of all genes identified by sequence, i.e. the determination of the role that the corresponding gene products play in plant metabolism or development. This functional genome analysis requires further new techniques and approaches that would allow one to establish causal relations between a certain DNA sequence and a function within a certain process. This could, in the most direct fashion, be provided by the identification of an altered phenotype caused by a mutation in a particular gene. One way of establishment of such a link is the map-based gene cloning mentioned earlier, which, in the future, will be greatly facilitated by the availability of the full genome sequence combined with the use of novel molecular markers based on the DNA-chip technology (Wang *et al.*, 1996) and the use of large DNA transfer systems such as the BiBAC (Hamilton *et al.*, 1996). Furthermore, the increasing number of lines carrying insertion mutations caused by the integration of the *Agrobacterium* T-DNA (e.g. Feldman, 1991) or derivatives of maize transposable elements (Osborne and Baker, 1995) will continue to be a highly useful resource to isolate the genes at the molecular level which were identified by a mutant phenotype. While most of the insertion mutagens used until now cause recessive mutations due to the "knock out" of the gene function by integration into the coding region or essential promoter regions, DNA fragments carrying a strong enhancer fragment could cause dominant mutations through the activation of genes located close to the insertion site ("activation tagging"; Walden *et al.*, 1994). The resulting ectopic expression, or the overexpression of the affected gene, could uncover the function of genes that may not be identified by loss of function mutations (e.g. in the cases of strict gametophytic lethality or in the case of genetic redundancy). In contrast to this "forward genetic" approach that proceeds from a mutant to the corresponding gene, the use of a large number of lines carrying insertions could also be used for a "reversed genetic" approach. Here, the previously determined sequence of a gene is used in combination with the known sequence of the inserted fragment to perform a search via PCR for insertions in the gene of interest. Applying a pooling strategy allows even tens of thousands of DNA probes (representing different lines carrying insertions at different positions within the genome) to be screened providing a good chance for the identification of a desired insertion (McKinney *et al.*, 1995). The insertion mutant thus isolated may be studied at the morphological and physiological level to uncover the function of the respective gene. An EC funded project (coordinated by A. Pereira, Wageningen, The Netherlands) is being conducted as a joint effort of eight labs to provide a saturation of the *Arabidopsis* genome with an assortment of T-DNA and transposon insertions. This methodology, however, is not limited to *Arabidopsis* for which already a world-wide collection of some 100,000 insertions has been generated, but is also extensively applied to maize. Here, lines carrying many copies of highly-active transposable elements of the *Mu* family have been selected. A collection of DNAs from approximately 40,000 maize plants each carrying multiple-inserted *Mu* family elements has been established by PIONEER HI-BRED INTERNATIONAL, INC. referred to as The TRAIT UTILITY SYTEM for CORN (TUSC). This saturated collection of insertions in the maize genome may also be useful to study genes originally isolated form rice in the frame of the corresponding genome project. A further reversed genetic approach, applicable to plants for which efficient genetic transformation procedures are available, is to reduce gene expression by antisense-inhibition or by cosuppression (e.g. Jorgensen, 1996). Usually in such experiments, a set of transgenic lines can be obtained that show different levels of gene inactivation and which are useful to correlate gene expression levels with phenotypic effects. Finally, overexpression and/or ectopic expression of genes (DNAs) placed under the control of heterologous promoters, like antisense-inhibition, advanced to a standard molecular genetic technique may be applied to the functional analysis of genes.

A resource of genetic variation present in the large collections of different germplasms available for many plant species has been used, until now only to a very limited extent for functional gene analysis (mostly for studies on disease resistance or stress tolerance). A major advantage of the genetic changes represented in these lines is the natural selection they were subjected to. This situation most probably allowed change of function, rather than loss of function, mutation to accumulate (in contrast to mutations induced through mutagenesis) the latter being less informative than the former. These resources will thus be of very high value, in particular as soon as the DNA-chip technology is developed further, to provide the basis for large scale comparative sequencing (using the available sequences as reference) at dimensions much larger than that possible with the current sequencing technology (Lander, 1996).

HOW MUCH CAN WE AFFORD IN PLANT GENOMICS AND HOW MUCH INFORMATION DO WE NEED

One important question with respect to the development of genomics in higher plants is as to how many plant species have to be investigated in the same depth as described for *Arabidopsis* and corn. Looking at the decent number of agriculturally-important plants it becomes obvious that such a detailed analysis is not affordable, at least with present technology.

However, there are many arguments that a very detailed study of two to three model species is probably enough for most applications of genomics, also in other species. This argument is based upon the two following considerations:

a) Genes which have been identified on the molecular level in one model species can easily be identified in other plant species by techniques such as heterologous hybridisation or PCR approaches.

b) In the case of map-based cloning approaches the high level of synteny observed between species such as all cereals or many dicotyledonous plants (cf. e.g. Moore, 1995; Paterson *et al*, 1996; Kurata *et al.*, 1994) allows the jumping between molecular markers of the organism of interest and corresponding markers of a model organism. Once two or three markers covering the region of interest in the crop species of interest have been correlated with markers in a model organisms and shown to be collinear in arrangement, it is a reasonable assumption that a significant amount of the DNA regions located between the markers will be collinear between both species too. Assuming that this region has been functionally and physically mapped in the model organism, this allows fast access to corresponding YACS or BACS clones of the organism of interest.

As a further extension of this scenario, once the entire genome of two or three model plants has been sequenced, and the sequences assigned to function, one could envision even a faster map-based cloning. In this approach, syntenic fragments covering the locus responsible for the phenotype in case of the crop species of interest would be compared to the fully sequenced and function-assigned corresponding fragment of the model organism. Educated guesses based on the biochemistry or physiology of the locus in question might allow the identification of only a few candidate genes. Final proof of the function has to be obtained by reversed genetics.

An analogous argument can be made for the functional genomics. Genomes saturated on the one hand by disruptive insertions and on the other hand by activator tags are necessary in two to three species in order to learn about loss of function and gain of function mutant phenotypes. Again, in most cases, this information can be used to look for similar genes in the other crop species, based on synteny or heterologous cloning approaches.

Thus in conclusion: in crop species not belonging to the extensively studied crops such as rice, corn, tomato and (though not a crop species) *Arabidopsis*, it is definitely necessary, but might also be sufficient, to have a dense map of molecular markers and a representative BACs or YACs library to allow exploitation of the full power of genomics, also in the less studied crop species.

CASE STUDIES

New approaches in creating male sterility systems for hybrid seed production

Hybrid seed offers major advantages to both the farmer and the breeder because of increased yield, high vigour and protection of the breeders germplasm. Up to now, only a few crops are sold predominantly as hybrid seed, most prominent amongst them, beeing corn. One major disadvantage, preventing the more widespread use of hybrid seed technology, is the associated high cost. The establishment of simple genetic ways using transgenic approaches, such as anther-specific expression of cell destructive principles, is therefore one possible solution and has been practised with great success.

Genomics allow the search for new approaches to solve this problem. More specifically, in a first approach, genes which, upon disruption, will only harm pollen development, must be identified. This can be done e.g. by saturation insertion mutagenesis or by chemical mutagenesis followed by map-based cloning approaches.

The identification of one or several genes in this way fulfils two different purposes:

a) a recessive locus has been identified giving rise to male sterility

b) upon coupling the coding region of the gene in question to a promoter the activity of which can be controlled externally (e.g. a promoter inducible by a chemical) male fertility can be restored at will, thus simplifying significantly the cost of maintaining the male-sterile line in a homozygous state.

Molecular identification of disease resistance genes

The molecular identification of disease-resistance genes represents probably one of the most breath-taking examples where the application of structural and functional genomics, in combination with transgene technology, has ultimately led to a most impressive achievement by the identification of genes conferring resistance to various pathogens (cf. e.g. Staskawicz et al., 1995 for a review).

It is interesting to note in this respect that both map-based cloning and insertion mutagenesis strategies have been used in the various groups. As a further consequence the insight into the structure of various resistance genes allowed the identification of structures conserved between them which immediately resulted in a new brute force cloning of a flood of resistance gene candidates by PCR approaches using degenerate primers (Kanazin et al., 1996; Yu et al., 1996; Leister et al., 1996). Thus, the field of disease resistance is open as never before, nearly exclusively due to the massive use of genomic approaches.

Identification of new herbicide targets

Herbicides are the largest class of compounds in crop protection. The development of new herbicides can be supported by genomics by, on the one hand, identifying via e.g. screening descendants of an insertion mutagenesis program for lethal events. Assuming that the lethal event is due to the inactivation of a certain enzyme/protein, it is plausible to assume that inactivation of the same protein by a chemical (which would be a herbicide) would also be lethal. Therefore, this protein represents a possible target for herbicides to be developed. Recloning the gene encoding this enzyme/protein and expressing it in a system allowing the set up of a high through put system for screening combinatorial libraries, allows the development of new chemicals with a herbicidal action, directed against new targets. Furthermore knowing the sequence and thus possibly, the function of this protein, allows a rapid decision as to whether this protein is plant-specific which might represent an advantage with respect to toxicological problems associated with the new herbicide.

Identification of new fungicide targets

Fungicides are the third largest class of agrochemicals. As described for the development of new herbicides, the identification of new targets for fungicides represents a major problem.

Again genomics might be used to help solving this problem. As the ideal fungicide, in addition to being safe from a toxicological and environmental standpoint, should be acting upon a target specific for the fungus, the approach of sequencing entire genomes or cDNAs of important phytopathogenic fungi seems attractive. By comparing the sequences and the putative functions encoded by the fungus and the genomic make-up of non-target organism, it should be possible to identify fungus-specific targets. A similar approach is presently already followed in case of bacterial diseases infectious for humans.

REFERENCES

Bevan, M., Ecker, J., Theologis, S., Federspiel, N., Davis, R., McCombie, D., Martienssen, R., Chen, E., Waterston, B., Wilson, R., Rounsley, S., Venter, C., Tabata, S., Salanoubat, M., Quetier, F., Cherry, M., Meinke, D. (1997) Objective: the complete sequence of a plant genome. *The Plant Cell* **9**, 476-478.

Botstein, D., White, R.L., Skolnick, M. and Davis, R.W. (1980) Construction of a genetic linkage map in man using restriction fragment length polymorphisms. *American Journal of Human Genetics* **32**, 314-31.

Delseny, M., Cooke, R., Raynal, M. and Grellet, F. (1997) The *Arabidopsis thaliana* cDNA sequencing projects. *FEBS Letters* **403**, 221-224.

Feldman, K.A. (1991) T-DNA insertion mutagenesis in *Arabidopsis*: mutational spectrum. *The Plant Journal* **1**, 71-82.

Giraudat, J., Hauge, B.M., Valon, C., Smalle, J., Parcy, F. and Goodman, H.M. (1992) Isolation of the *Arabidopsis ABI3* gene by positional cloning. *Plant Cell* **4**, 1251-1261.

Goodman, H.M., Ecker, J. and Dean, C. (1995) The genome of *Arabidopsis thaliana*. *Proceedings of the National Academy of Sciences USA* **92**, 10831-10835.

Hamilton, C.M., Frary, A., Lewis, C. and Tanksley, S.D. (1996) Stable transfer of intact high molecular weight DNA into plant chromosomes. *Proceedings of the National Academy of Sciences USA* **93**, 9975-9979.

Hearne, C.M., Ghosh, S. and Todd, J.A. (1992) Microsatellites for linkage analysis of genetic traits. *Trends in Genetics* **8**, 288-294.

Jorgensen, R.A., Cluster, P.D., English, J., Que, Q., Napoli, C.A. (1996) Chalcone synthase cosuppression phenotypes in petunia flowers: Comparison of sense vs. antisense constructs and single-copy vs. complex T-DNA sequences. *Plant Molecular Biology* **31**, 957-973.

Kanazin, V., Marek, L. and Shoemakert, R. (1996) Resistance gene analogs are conserved and clustered in soybean. *Proceedings of the National Academy of Science USA* **93**, 11746 - 11750

Konieczny, A. and Ausubel, F.M. (1993) A procedure for mapping *Arabidopsis* mutations using co-dominant ecotype-specific PCR-based markers. *The Plant Journal* **4**, 403-410.

Kurata, N., Moore, G., Nagamura, Y., Foote, T., Yano, M., Minobe, Y. and Gale, M. (1994) Conservation of genome structure between rice and wheat. *Bio/Technology* **12**, 276 - 278

Lander, E. (1996) The new genomics: global views of biology. *Science* **274**, 536-539.

Leister, D., Ballvora, A., Salamini, F. and Gebhardt, C. (1996) A PCR-based approach for isolating pathogen resistance genes from potato with potential for wide application in plants. *Nature Genetics* **14**, 421 - 429

Martin, G.B., Brommonschenkel, S.H., Chunwongse, J., Frary, A., Ganal, M.W., Spivey, R., Wu, T., Earle, E.D. and Tanksley, S. (1993) Map-based cloning of a protein kinase gene conferring disease resistance in tomato. *Science* **262**, 1432-1436.

McKinney, E.C., Aali, N., Traut, A., Feldmann, K.A., Belostotsky, D.A., McDowell, J.M., and Meagher, R.B. (1995) Sequence.based identification of T-DNA insertion mutations in *Arabidopsis*: actin mutants *act2-1* and *act4-1*. *The Plant Journal* **8**, 613-622.

Moore, G. (1995) Cereal genome evolution : Pastoral pursuits with 'Lego 'genomes. *Current Opinion in Genetics and Development* **5**, 717 - 724

Mozo, T., Birren, B., Lehrach, H., Meier-Ewert, S., Meyerowitz, E., Shizuya, H., Simon, M., Willmitzer, L. and Altmann, T. (1996) Development of a bacterial artificial chromosome library for *Arabidopsis thaliana* and its use for physical mapping. The 7th International Conference on *Arabidopsis* Research, Norwich, UK, June 23 - 27 1996; Abstract P130

Newman, T., de Bruijn, F.J., Green, P., Keegstra, K., Kende, H., McIntosh, L., Ohlrogge, J., Raikhel, N., Somerville, S., Tomashaw, M., Retzel, E. and Somerville, C. (1994) Genes Galore: A summary of methods for accessing results from large-scale partial sequencing of anonymous *Arabidopsis* cDNA clones. *Plant Physiology* **106**, 1241-1255.

Osborne, B.J. and Baker, B. (1995) Movers and shakers: Maize transposons as tools for analyzing other plant genomes. *Current Opinion in Cell Biology* **7**, 406-413.

Paterson, A., Lan, T., Reischmann, K., Chang, C., Lin, Y., Liu, S., Burow, M., Kowalski, S., Katsar, C., DelMonte, T., Feldmann, K., Schertz, K. and Wendel, J. (1996). Toward a unified genetic map of higher plants, transcending the monocot-dicot divergence. *Nature Genetics* **14**, 380 - 382

Putterill, J., Robson, F., Lee, K., Simon, R. and Coupland, G. (1995) The *CONSTANS* gene of *Arabidopsis* promotes flowering and encodes a protein showing similarities to zinc finger transcription factors. *Cell* **80**, 847-857.

Schmidt, R., West, J., Love, K., Lenehan, Z., Lister, C., Thompson, H., Bouchez, D. and Dean, C. (1995) Physical map and organization of *Arabidopsis thaliana* chromosome 4. *Science* **270**, 480-483.

Staskawicz, B.J., Ausubel, F.M., Baker, B.J., Ellis, J.G. and Jones, D.G. (1995) Molecular genetics of plant disease resistance. *Science* **268**, 661 - 667

Tanksley, S.D. (1993) Mapping polygenes. *Annual Review of Genetics* **27**, 205-233.

Tanksley, S.D., Ganal, M.W. and Martin, G.B. (1995) Chromosme landing: a paradigm for map-based gene cloning in plants with large genomes. *Trends in Genetics* **11**, 63-68.

Vos, P., Hogers, R., Bleeker, M., Reijans, M., Van De Lee, T., Hornes, M., Frijters, A., Pot, J., Peleman, J., Kuiper, M. and Zabeau, M. (1995) AFLP: A new technique for DNA fingerprinting. *Nucleic Acids Research* **23**, 4407-4414

Walden, R., Fritze, K., Hayashi, H., Miklashevichs, E., Harling, H. and Schell, J. (1994) Activation tagging: a means of isolating gene implicated as playing a role in plant growth and development. *Plant Molecular Biology* **26**, 1521-1528.

Wang, D., Sapolsky, R., Spencer, J., Rioux, J., Kruglyak, L., Hubbell, E., Ghandour, G., Hawkins, T., Hudson, T., Lipshutz, R. and Lander, E. (1996) Toward a third generation genetic map of the human genome based in bi-allelic polymorphisms. 46th Annual Meeting of the American Society of Human Genetics, San Francisco,

California, USA, October 29-November 2, 1996. *American Journal of Human Genetics* **59** A3.

Wang, G.-L., Holsten, T.E., Song, W., Wang, H.-P. and Ronald, P. (1995) Construction of a rice bacterial artificial chromosome library and identification of clones linked to a disease resistance locus. *The Plant Journal* **7**, 525-533.

Williams, J.G.K., Kubelik, A.R., Livak, K.J., Rafalski, J.A. and Tingey, S.V. (1990) DNA polymorphisms amplified by arbitrary primers are useful as genetic markers. *Nucleic Acids Research* **18**, 6531-6536.

Yu,Y., Buss, G. and Maroof, M. (1996) Isolation of a superfamily of candidate disease-resistance genes in soybean based on a conserved nucleotide-binding site. *Proceedings of the National Academy of Science USA* **93**, 11751 - 11756

Zachgo, E.A., Wang, M.L., Dewney, J., Bouchez, D., Camilleri, C., Belmonte, S., Huang, L., Dolan, M. and Goodman, H. (1996) A physical map of chromosome 2 of *Arabidopsis thaliana*. *Genome Research* **6**, 19-25.

Zhang, H.-B., Choi, S., Woo, S.-S., Li, Z. and Wing, R. (1996) Construction and characterization of two rice bacterial artificial chromosome libraries from parents of a permanent recombinant inbred mapping population. *Molecular Breeding* **2**, 11-24.

THE USE OF GENE SEQUENCES IN COMBATING PESTICIDE RESISTANCE

B. J. MIFLIN, L.M. FIELD, M.S. WILLIAMSON, A.L. DEVONSHIRE and I. DENHOLM

IACR Rothamsted, Harpenden, Herts, UK.

INTRODUCTION

The genomes of major insect and other pests are not yet being sequenced *in toto* and thus the 'genomics' revolution has not yet played a part in understanding and combating insecticide resistance. Nevertheless, genetic technologies and knowledge of gene sequences have played a major part in our current understanding of how weeds, insects and fungi resist the action of toxic chemicals, and how we can devise management strategies to alleviate and prevent resistance from frustrating their control. This can be illustrated by showing how genetic techniques, from the use of markers to full isolation and characterisation of the mutations present in target sites, have illuminated our knowledge of the mechanisms of resistance.

APPLICATIONS OF MARKER TECHNOLOGY

In most cases, when a pest first shows resistance to an insecticide, the nature of the mechanisms involved are not known. In the past, the answer was eventually found by a series of trial and error investigations in which a number of educated guesses were studied in some detail, until a suitable body of evidence could be assembled. With the advent of a whole range of genetic marker technology involving RFLPs, AFLPs, RAPDS and microsatellites, it is possible to map the pest genome and to locate the loci responsible for resistance. This approach is equally applicable to pest species for which little genetic information was previously available, as it is to well-studied model systems such as *Drosophila melanogaster*. It is particularly valuable where resistance is due to a number of loci, since it allows all of the major ones to be identified and to allocate some relative value to their importance. Knowing the position of resistance loci can also be used to correlate with the position of potential candidate genes on the map and, if the genes have not been isolated, as the start for map-based cloning of the loci.

A current example of where this approach is proving very valuable in trying to address an imminent problem was recently given by Heckel *et al.* (1997). Genes for the insecticidal toxins of *Bacillus thuringiensis* (Bt) have been introduced into a number of crops which are now being introduced into agriculture in the USA and Australia. These releases are expected to increase greatly the selection pressure on the insects and resistance may be expected to occur. Strains of diamondback moth (*Plutella xylostella*) and Indianmeal moth (*Plodia interpunctella*), resistant to Bt toxins, have appeared in the field where spore-based Bt insecticides have been used over many years, and other species have been selected for Bt resistance in the laboratory (Tabashnik, 1994). The strain of tobacco budworm (*Heliothis virescens*) selected by the Monsanto Company was 20-fold resistant to the toxin Cry1Ac (Stone *et al.*, 1989), and Heckel has shown that this is due to the action of at least three distinct linkage groups (D. G. Heckel, pers. comm., 1997). However, the most resistant strain (Gould *et al.*, 1995) is up to 10,000-fold resistant to this toxin. Genetic analysis shows that 80% of this resistance is due to a recessive gene at locus *BtR-4* on linkage group 9. There are many possible mechanisms that might be encoded by *BtR-4* and it is, therefore, of high priority to clone this locus, an endeavour that will be greatly aided by marker technology, which is under way (Heckel *et al.*, 1997).

THE ROLE OF GENOME AMPLIFICATION

Gene amplification has long been documented as a means of resistance to insecticides and other xenobiotics (Devonshire and Field, 1991). *Myzus persicae,* and its tobacco feeding form *M. nicotianae,* have developed resistance in this way to organophosphate, carbamate and pyrethroid insecticides. The first mechanism recognised was the overproduction of one of two closely related carboxylesterases (E4 or FE4) that degrade and sequester insecticidal esters, and it has been shown that both species have the same amplified E4 or FE4 genes (Field *et al.*, 1994). These amplified genes are present in clusters inserted in the chromosomes and are not present in separate mini-chromosomes as commonly occurs in cell cultures resistant to cytotoxic drugs. Studies of the E4 amplification events show that aphids of wide geographic origin have the amplified genes on identical repeat units of DNA, suggesting that the mechanism arose by a single event that has spread by migration. This may be related to the apparently complete linkage between amplification of E4 and a chromosomal translocation (Blackman *et al.*, 1978). This translocation appears to impose some reproductive isolation on clones as they do not produce overwintering sexual forms, or only males, and are permanently pathenogenic.

Recent surveys of aphids from around the world (Field *et al.*, 1997) show that high levels of resistance are the result of both E4 and FE4 amplification, and that the amplified genes are methylated at *Msp*I sites. Surveys have also found that there are insects which have amplified E4 genes but no resistance (i.e. they are apparent revertants). Further analysis has shown that the amplified E4 sequences of such revertants lose their methylation at *Msp*I sites during reversion to the susceptible phenotype (Hick *et al.*, 1996).

A key question in resistance development and management is whether or not there is a fitness disadvantage associated with the resistance mechanism. Experiments comparing the over-wintering success of clones of known E4 or FE4 phenotypes demonstrated that, for the coldest and wettest months in particular, there is an inverse relationship between survival in Britain and the level of esterase gene amplification (Foster *et al.*, 1996). This could explain, at least partially, the temporal stability of the esterase phenotypes that has been observed over several seasons.

```
120E3T7     1  CAAACCAATATCAAAAGAAACCAAAAGCAATCTTTCAAAAATGATTT  47
               ||||||||||||||||||||||||||||||||||||||||||||||
E4cDNA   1504  CAAACCAATATCAAAAGAAACCAAAAGCAATCTTTCAAAAATGATTT  1550

120E3T7        CAGATAGATCATTTGGATATGGTACAAGTAAAGCAGCCCAGCATATAGCT  97
               |||||||||||||||||||||||||||||||||||||||||||||||||
E4cDNA         CAGATAGATCATTTGGATATGGTACAAGTAAAGCAGCCCAGCATATAGCT  1600

120E3T7        GCAAAGAATACGGCACCTGTATATTTCTATGAATTTGGCTACAGTGGTAA  147
               |||||||||||||||||||||||||||||||||||||||||||||||||
E4cDNA         GCAAAGAATACGGCACCTGTATATTTCTATGAATTTGGCTACAGTGGTAA  1650

120E3T7        TTATTCTTACGTAGCTTTTTTCGATCCNAAATCATATTCCCGGGGTTCAA  197
               ||||||||||||||||||||||||||:|||||||||||||||||||||||
E4cDNA         TTATTCTTACGTAGCTTTTTTCGATCCAAAATCATATTCCCGGGGTTCAA  1700

120E3T7        GCCCGACTCATGGCGATGAAACCAGCTATGTATTAAAAATGGATGGTTTC  247
               ||||||||||||||||||||||||||||||||||||||||||||||||||
E4cDNA         GCCCGACTCATGGCGATGAAACCAGCTATGTATTAAAAATGGATGGTTTC  1750

120E3T7        TACGTTTACGACAATGANGAAGATAGAAAGATGATCAAAACTATGGGTTA  297
               |||||||||||||||||:|||||||||||||||||||||||||||||  |
E4cDNA         TACGTTTACGACAATGAAGAAGATAGAAAGATGATCAAAACTATGG..TT  1798

120E3T7        ATATTTTGGGCAACTTTTTNTCCAAATCTGGGNGTNCCAGATACCTGAAA  347
               |    ||||||||||||||| :    || :   ||:||:|||||||  ||
E4cDNA         AATATTTGGGCAACTTTTAT...CAAATCTGGGAGTACCAGATAC..TGAA  1843

120E3T7        ATTTCAGGAATTTTGGGTACCCGGTTTCTAAGGNTCTAGCAGATCCTTTC  397
               |  ||||| ||||| |||  || |||||||||| :|||||||||||||||
E4cDNA         AATTCAGAAATTT..GGTTACCTGTTTCTAAGAATCTAGCAGATCCTTTC  1891

120E3T7        AGGTTCACTAAGGTTTCTCAACAACAAACATTTTTNNGCCNGGGGGCANT  447
               ||||||||||| || ||||||||||||| ||| ::|||:| |  ||:|
E4cDNA         AGGTTCACTAAGATTACTCAACAACAAACA.TTTGAAGCCAGAGAACAAT  1940

120E3T7        TTNACCCCGGGNTTTNTNATTTTGGGGGNGGNTTCCCCNAATT.......  490
                : |  | ||  :||  : :||||||| | :| :| ||  :|| |
E4cDNA         CAACCACGGGAATTATGAATTTTGGAGTAGCTTACCATTAAATGAATTTT  1990
```

FIGURE 1. Comparison of the sequence of esterase E4 and an unknown EST from an *Arabidopsis thaliana* library.

120E3T7 = 490 bp sequence from *A. thaliana* cDNA clone 120E3T7 (Newman *et al.*, 1994). Database Accession number T43946.

E4cDNA = cDNA sequence of E4 esterase from *Myzus persicae* (Field *et al.*, 1994). Database Accession number X74554.

THE ROLE OF INSECTICIDE RESISTANCE IN GENOMIC RESEARCH

The title of this section has been chosen to serve as a warning of the dangers that can occur in carrying out studies on the genomes of infected organisms. In the EST library of *Arabidopsis* there is a clone 120E3T7 (Newman *et al.*, 1994). When this sequence is compared to the E4 esterase of *M. persicae* there is a striking sequence match (Figure 1). One interpretation would be that the plant also has the identical E4 esterase gene to *M. persicae*, but a more likely conclusion is that the cDNA library was prepared from plants contaminated by aphids, which were highly insecticide resistant due to the presence and expression of an amplified esterase gene. Small levels of aphid contamination could in this case have given rise to sufficient RNA for the corresponding cDNA to appear in the EST library. How many other such examples are there in libraries around the world?

SEQUENCE VARIATION IN TARGET SITES

Resistance based on altered target sites is well-established across all pests. Where sequences of the gene for that site are known, it has been possible to find mutations that can explain the resistance. Depending on the nature of the target site and its interaction with the natural substrate, there may be only one mutation responsible or there may be many. Herbicide target sites provide examples from both ends of the scale (see Mazur and Falco (1989) for review). At one end is the herbicide glyphosate which inhibits the enzyme 5-enolpyruvylshikimic acid-3-phosphate synthase (EPSPS) that catalyses the important first step in aromatic amino acid biosynthesis in plants (Amrhein *et al.*, 1980). Despite being widely used over many years, there is probably only one case of resistance arising in the field, and it is not proven that this is due to any alteration in the enzyme. Laboratory studies with plant tissue cultures failed to give rise to any mutant forms of the enzyme, even under high selective pressure, but some tolerance was provided by gene amplification (Steinrucken *et al.*, 1986); a finding that has not been duplicated in the field. By cloning the *AroA* gene, which encodes EPSPS from *Salmonella*, and inserting it into tobacco, Comai and colleagues were able to demonstrate a measure of glyphosate tolerance (Comai and Stalker, 1986). EPSPS genes from many bacterial sources have been tried (see Mazur and Falco, 1989). Currently, glyphosate-tolerant crops are being generated using the EPSP gene from an *Agrobacterium* sp strain CP4, that is highly effective at degrading glyphosate. Additional glyphosate-tolerance can be obtained by transformation of crops with the gene for glyphosate oxidoreductase from *Achromobacter* (Wells, 1995). In line with these studies, which suggest a very limited opportunity to discriminate between the pesticide and the natural substrate of the enzyme, no other analogues of glyphosate, nor any other classes of chemical, have been found which are as effective inhibitors of EPSPS.

In contrast, the enzyme acetolactate synthase, a key enzyme in branched chain amino acid synthesis in plants (Bryan, 1990) has been found to be susceptible to at least four different classes of herbicide chemistry and, within one class, for example the sulfonylureas there are many compounds that are effective. This has been ascribed to the binding of the compounds to a vestigial ubiquinone binding site that is no longer required for enzyme function, but such binding might lead to steric interference of binding of substrate and enzyme (Schloss, 1990). This wide range of susceptibility is also correlated with the relative ease with which resistance is found to these herbicides and the large number of mutations in the DNA sequence of the gene which are associated with resistance (Mazur and Falco, 1989).

Insecticide resistance has also been shown to be due to mutations in genes encoding target sites. The enzyme acetylcholinesterase (AChE) is the target site for organophosphate and carbamate insecticides and, as a response to selection, a number of insect species have developed forms of AChE that are much less sensitive to inhibition (Brattsten, 1990). Isolation of the wild-type gene and its variants from resistant insects has allowed the amino acid substitutions that confer reduced sensitivity to be identified (Mutero *et al.*, 1994). Knowledge of how these substitutions might affect both substrate and insecticide interaction with AChE has been aided by the determination of the crystal structure of this enzyme from the electric fish, *Torpedo californica* (Sussman *et al.*, 1991). Until recently, this mechanism was not thought to be important in resistance in *M. persicae*, but an insensitive form of AChE which confers resistance specifically to the carbamate pirimicarb and to triazamate, a novel triazole aphicide, has now been found (Moores *et al.*, 1994). This is of considerable concern, because pirimicarb has long been the favoured aphicide for controlling resistant aphids with elevated esterase (Field *et al.*, 1994).

The primary physiological target for DDT and pyrethroid insecticides is the voltage-sensitive sodium channel of nerve membranes. An important mechanism that confers resistance to both classes of insecticide is termed knockdown resistance (*kdr*) and is characterised by reduced sensitivity of the insect nervous system. Strains of housefly with greatly enhanced resistance have also been found and their resistance termed *super-kdr*. Evidence that the resistance of *kdr* insects results from a modification of the target site, came from studies of known sodium channel neurotoxins and binding studies of pyrethroids to the sodium channel of *super-kdr* houseflies (reviewed in Williamson *et al.*, 1996a). The discovery of two sodium channel genes in *Drosophila* provided DNA sequences which could be used to map resistance loci in several species (see Heckel *et al.*, 1997; Williamson *et al.*, 1996b) and the results showed that there was close linkage between *kdr* resistance and the *para*-type sodium channel gene. This gene has been cloned from the housefly and comparative sequencing studies done on the wild-type, *kdr* and *super-kdr* alleles (Williamson *et al.*, 1996b). The conclusion from these studies is that there are two mutations in the part of the gene that encodes that portion of the protein that provides the intracellular 'mouth' of the channel pore; these are a mutation of leucine in position 1014 to phenylalanine (found in *kdr* and *super-kdr* strains) and the mutation of the methionine to threonine at position 918 (found only in *super-kdr* strains). The *kdr* mutation of Leu to Phe has recently been found in *M. persicae* and is also associated with resistance to pyrethroids and DDT (Martinez-Torres *et al.*, 1997).

COMPARATIVE GENETICS

Genomic research in other species can provide many useful tools for understanding resistance in pests, exemplified by the *para*-sodium channel gene from *Drosophila*. As complete sequences of organisms related to pest species become available, they will be extremely useful. In plants, the sequencing of the *Arabidopsis* and rice genomes will provide tools for probing resistance mechanisms in dicotyledonous and monocotyledonous weeds and may also provide leads for identifying selective herbicidally-active chemicals. One example where comparative genetic information from a number of different species is proving useful is that for the enzyme acetylCoA carboxylase (ACCase). This enzyme is important in the biosynthesis of fatty acids in plants, fungi and bacteria. Each of these organisms has multiple forms including cytoplasmic and organellar isoenzymes. The cytoplasmic and plastidic forms in wheat and other Poaceae are both of the eukaryotic type and are sensitive to the herbicides of the aryloxyphenoxy propionate and cyclohexanedione type, whereas the plastidic form in dicotyledons is of the prokaryotic type and insensitive (Konishi and Saskaki, 1994). The experimental fungicide soraphen inhibits the enzyme from fungi and rat liver mitochondria, but not the plastidic form from spinach (Pridzun *et al.*, 1995). Most of the cDNAs and genes for these enzymes from different species have been sequenced, which should be useful in the search for more differentially active chemicals, and in understanding the development of resistance to them (see Podkowinski *et al.*, 1996).

Comparative genetics are also important in determining whether the conclusions that can be drawn from determination of resistance mechanisms in one species can be applied to others. Recently Martinez-Torres *et al.* (1997) surveyed a range of insect species to determine the sequence of the *para*-sodium channel gene around amino acids positions 918 and 1014. All susceptible insects from four orders contained Met and Leu in the respective positions. Pyrethroid-resistant strains of *P. xylostella* and *M. persicae* both contained the housefly *kdr* Leu to Phe mutation. Similarly, a single amino acid change from alanine to serine in the target protein for cyclodiene insecticides (a GABA-gated chloride channel) has been shown to occur in a diverse range of insect pests species (Thompson *et al.*, 1993).

MULTIPLE MECHANISMS

Pests with broad spectrum resistance to more than one chemical class are becoming more wide-spread. In some cases, such as in the weeds *Lolium rigidum* and *Alopecurus myosuroides,* there is evidence for broad non-specific mechanisms for herbicide degradation (Hall *et al.*, 1994). In *M. persicae* there is now detailed evidence that a number of mechanisms can be present in a single strain of aphid. It is only with the advent of sequence-specific techniques that this has been fully understood. Surveys of 58 *M. persicae* clones from around the world showed that all of the above mechanisms of insecticide resistance were present (Field *et al.*, 1997). Resistant AChE phenotypes were found in eleven clones, all of which also had elevated levels of non-specific esterase (E4 or FE4). Although it had been thought originally that elevated esterase was providing field resistance against pyrethroids and DDT, molecular analysis showed the presence of the *kdr* mutation in some of these aphids (Martinez-Torres *et al.*, 1997) and a DDT discrimination assay has been developed which correlates well with the *kdr* mutation (Field *et al.*, 1997). The 25 DDT-resistant clones all had amplified E4 genes, thus demonstrating strong linkage disequilibrium between these mechanisms. In all six clones had all three resistance mechanisms. Knowledge of the presence of different resistance mechanisms can assist the choice of insecticides (Table 1). However, their use has to be managed carefully to prevent further build up of resistance.

TABLE 1. Practical implications of resistance in *Myzus persicae*

Insecticide	Resistance mechanism		
	E4/FE4	MACE*	Kdr
Pirimicarb	+	-	+ +
Other carbamates	+	+ +	+ +
Triazamate	+	-	+ +
Organophosphates	+	+ +	+ +
Pyrethroids	+	+ +	-
Chloronicotinyls	+ +	+ +	+ +

+ + Good control;   + Reduced control;   - No control
*Modified AChE

CONCLUSIONS

The advent of recombinant DNA technologies has enhanced our ability to understand resistance to pesticides tremendously. The advent of a range of marker technologies has allowed the localisation of genes contributing to the resistance and, where sequences of the likely resistance genes do not exist, can lead to cloning of the resistance loci. Where the resistance genes are known and cloned, the mutations conferring resistance can be determined. In turn, these provide opportunities for PCR-based assays that can determine the distribution of the resistance genes within pest populations, even in the presence of multiple resistance mechanisms. Such information can be of practical use in the deployment of pesticides to combat or overcome resistance.

ACKNOWLEDGEMENTS

IACR receives grant-aided support from the Biotechnology and Biological Sciences Research Council of the UK. We are grateful to Dr David Heckel for information on his work prior to its publication.

REFERENCES

Amrhein, N., Deus, B., Gehrke, P. and Steinrucken, H.C. (1980) The site of the inhibition of the shikimate pathway by glyphosate. II Interference of glyphosate with chorismate formation *in vitro* and *in vivo*. *Plant Physiology*, **66**, 830-834.

Blackman, R.L., Takada, H. and Kawakami, K. (1978) Chromosomal rearrangement involved in the insecticide resistance of *Myzus persicae*. *Nature*, **271**, 450-452.

Brattsten, L.B. (1990) Resistance mechanisms to carbamate and organophosphate insecticides. In: *Managing Resistance to Agrochemicals*, M.B. Green, H.M. LeBaron and W.K. Moberg, (Eds) American Chemical Society, Washington, D.C. pp. 42-60.

Bryan, J.K. (1990) Advances in the biochemistry of amino acid biosynthesis. In: *The Biochemistry of Plants*, Vol. 16. B.J. Miflin and P.J. Lea (Eds), Academic Press, pp. 161-195.

Comai, L. and Stalker, D. (1986) Mechanisms of herbicide resistance and their manipulation by molecular biology. *Oxford Surveys of Plant Molecular and Cell Biology*, **3**, 166-195.

Devonshire, A.L. and Field, L.M. (1991). Gene amplification and insecticide resistance. *Annual Review of Entomology*, **36**, 1-23.

Field, L.M., Anderson, A.P., Denholm, I., Foster, S.P., Harling, Z.K., Javed, N., Martinez-Torres, D., Moores, G.D., Williamson, M.S. and Devonshire, A.L. (1997) Use of biochemical and DNA diagnostics for characterising multiple mechanisms of insecticide resistance in the peach-potato aphid, *Myzus persicae* (Sulzer). *Pesticide Science, * **51**, 283-289.

Field, L.M., Javed, N.J., Stribley, M.F. and Devonshire, A.L. (1994) The peach-potato aphid *Myzus persicae* and the tobacco aphid *Myzus nicotianae* have the same esterase-based mechanism of insecticide resistance. *Insect Molecular Biology*, **3**, 143-148.

Foster, S.P., Harrington, R., Devonshire, A.L., Denholm, I., Devine, G.J., Kenward, M.G. and Bale, J.S. (1996) Comparative survival of insecticide-susceptible and resistant peach-potato aphids *Myzus persicae* (Sulzer) (Hempiptera: Aphididae) in low temperature field trials. *Bulletin Entomological Research*, **86**, 17-27.

Gould, F., Anderson, A., Reynolds., A., Bumgarner, L., and Moar, W. (1995) Selection and genetic analysis of a *Heliothis virescens* (Lepidoptera: Noctuidae) strain with high levels of resistance to *Bacillus thuringiensis* toxins. *Journal Economic Entomology*, **88**, 1545-1559.

Hall, L.M., Holtum, J.A.M. and Powles, S.B. (1994) Mechanisms responsible for cross resistance and multiple resistance. In: *Herbicide Resistance in Plants: Biology and Biochemistry*, S.B. Powles and J.A.M. Holtum (Eds), CRC Press, Boca Raton, Florida, pp. 243-261.

Heckel, D.G., Gahan, L.C., Gould, F., Daly, J.C. and Trowell, S. (1997) Genetics of *Heliothis* and *Helicoverpa* resistance to chemical insecticides and to *Bacillus thuringiensis*. *Pesticide Science*, **51**, 251-258.

Hick, C.A., Field, L.M., and Devonshire, A.L. (1996) Changes in the methylation of amplified esterase DNA during loss and reselection of insecticide resistance in the peach-potato aphid *Myzus persicae. Insect Biochemistry and Molecular Biology*, **26**, 41-47.

Konishi, T., and Sasaki, Y. (1994) Compartmentalisation of two forms of acetyl-CoA carboxylase in plants and the origin of their tolerance towards herbicides. *Proceedings National Academy of Sciences USA*, **91**, 3598-3601.

Martinez-Torres, D., Devonshire, A.L., and Williamson, M.S. (1997) Molecular studies of knockdown resistance to pyrethroids: cloning of domain II sodium channel gene sequences from insects. *Pesticide Science*, **51**, 265-270.

Mazur, B.J. and Falco, S.C. (1989) The development of herbicide resistant crops. *Annual Review Plant Physiology*, **40**, 441-470.

Moores, G.D., Denholm I.H. and Devonshire, A.L. (1994) Insecticide-insensitive acetylcholinesterase can enhance esterase-based resistance in *Myzus persicae* and *Myzus nicotianae. Pesticide Biochemistry and Physiology*, **49** 114-120.

Mutero, A., Pravavorio, M., Bride, J.M. and Fournier, D. (1994). Resistance associated point mutations in insecticide-insensitive acetylcholinesterase. *Proceedings National Academy of Sciences USA*, **91** 5922-5926.

Newman, T., de Bruijn, F.J., Green, P., Keegstra, K., Kende, H., McIntosh, L., Ohlrogge, J., Raikhel, N., Somerville, S., Thomashow, M., Retzel, E., and Somerville, C. (1994) Genes Galore: A summary of methods for accessing results from large-scale partial sequencing of anonymous *Arabidopsis* cDNA clones. *Plant Physiology*, **106**, 1241-1255.

Podkowinski, J., Sroga, G.E., Haselkorn, R. and Gornicki, P. (1996) Structure of a gene encoding a cytosolic acetyl-CoA carboxylase of hexaploid wheat. *Proceedings National Academy of Sciences USA*, **93**, 1870-1874.

Pridzun, L., Sasse, F. and Reichenbach, H. (1995) Inhibition of fungal acetyl-CoA carboxylase: a novel target discovered with the myxobacterial compound soraphen: *Antifungal Agents: Discovery and Mode of Action*, G.K. Dixon, L.G. Copping, D.W. Hollomon (Eds). Bios Scientific Publishers, Oxford, pp. 99-109.

Schloss, J.V. (1990) Acetolactate synthase, mechanism of action and its herbicide binding site. *Pesticide Science,* **29**, 283-292.

Steinrucken, H.C., Schulz, A., Amrhein, N., Porter, C.A. and Fraley, R.T. (1986) Overproduction of a 5-enolpyruvyl-shikimic acid-3-phosphate synthase. *Biochemical & Biophysical Research Communications*, **94**, 1207-1212.

Stone, T.B., Sims, T.R. and Marrone, P.G. (1989). Selection of tobacco budworm for resistance to a genetically engineered *Pseudomonas fluorescens* containing the delta-endotoxin of *Bacillus thuringiensis* subsp. *kurstaki. Journal of Invertebrate Pathology*, **53**, 228-234.

Sussman J.L., Harel, M., Frolow, F., Oefner, C., Goldman, A., Toker, L. & Silman, I. (1991). Atomic structure of acetylcholinesterase from *Torpedo californica*: A prototypic acetylcholine-binding protein. *Science,* **253**, 872-879.

Tabashnik, B.E. (1994) Evolution of resistance to *Bacillus thuringiensis. Annual Review of Entomology,* **39**, 47-79.

Thompson, M., Steichen, J.C. & ffrench-Constant, R.H. (1993). Conservation of cyclodiene insecticide resistance-associated mutations in insects. *Insect Molecular Biology,* **2**, 149-152.

Wells, B.H. (1995) Development of glyphosate tolerant crops into the market. *Brighton Crop Protection Conference - Weeds 1995,* 787-790

Williamson, M.S., Martinez-Torres, D., Hick, C.A., Castells, N. and Devonshire, A.L. (1996a) Analysis of sodium channel gene sequences in pyrethroid-resistance houseflies. Progress towards a molecular diagnostic for knockdown resistance. In: *Molecular Genetics and Evolution of Pesticide Resistance.* T.M. Brown, (Ed) *ACS Symposium Series 645,* American Chemical Society, Washington, DC, USA, pp. 52-61.

Williamson, M.S., Martinez-Torres, D., Hick, C.A. and Devonshire, A.L. (1996b) Identification of mutations in the housefly *para*-type sodium channel gene associated with knockdown resistance (*kdr*) to pyrethroid insecticides. *Molecular and General Genetics,* **252**, 51-60.

IMPACT OF GENOMICS ON IMPROVING THE QUALITY OF AGRICULTURAL PRODUCTS

D.J. MURPHY

John Innes Centre, Norwich Research Park, Colney Lane, Norwich, NR4 7UH, UK

INTRODUCTION

Agriculture provides a renewable supply of the vast majority of our foodstuffs and of many valuable non-food products, including oleochemicals, polymers, cosmetics and even pharmaceuticals. The application of scientific plant-breeding techniques during the 20th century has led to enormous improvements in both yield and quality of agricultural products, which has permitted the enormous growth in the human population of the planet. Nevertheless, conventional plant-breeding techniques are limited to existing genetic variation within individual species, or sexually compatible related species.

The past decade has witnessed dramatic developments in the potential for the further manipulation of major agricultural crops, using biotechnological methods based largely on an improved knowledge of genomics. These developments now allow for the introgression of agriculturally useful traits from widely related species, using techniques such as protoplast fusion and embryo rescue, in conjunction with powerful molecular marker-assisted selection based on the use of RFLP (restriction fragment length polymorphism), AFLP (amplified fragment length polymorphism) and microsatellite markers. The advent of genetic engineering has now widened still further the scope for creating additional genetic variation, so that genes may now be accessed from any biological organism, or indeed made synthetically, and then transferred to the crop species of interest. These methodologies will also allow for the much more rapid domestication of new crop species, producing valuable edible or industrial products, within a timescale of decades rather than centuries. In this review, some of the recent scientific advances which underpin the application of genomics to the improvement of quality traits in both existing and potential crop species will be examined.

DESIRABLE QUALITY TRAITS IN CROPS

The quality of the products produced from crop plants is determined largely by their chemical compositions. For example, the nutritional quality of many seed crops is affected by the amino acid composition of the proteins produced by such seeds. Seed proteins tend to be rich in amino acids such as glutamine and asparagine and are often relatively deficient in sulphur-containing amino acids such as methionine and cysteine. Several amino acids such as lysine and methionine are essential for human and animal nutrition and deficiencies can have serious effects on physiological development. Indeed, the over-enthusiastic adoption of vegetarian diets has led to amino acid-deficiency symptoms in human infants. One of the objectives of crop researchers is, therefore, to improve the essential amino acid content of seed proteins.

Many crops produce organs which are particularly enriched in starches, although the composition of starch in crops such as potato is quite different from that of wheat or rice. These differences in starch structure do not have a strong bearing on its suitability for either human or animal nutrition, but are very important in the use of starches in processed foods or for industrial products (Smith & Martin, 1993; Batchelor *et al.*, 1996). For example, potato starch, with its large granule size and high degree of phosphorylation, is particularly suited for paper manufacture, while the *waxy* variety of maize produces a low amylose starch which is useful in the manufacture of processed foods for freezing.

Numerous seed and fruit crops are grown as sources of vegetable oils. The quality of a given vegetable oil is dictated by its fatty acid composition. Several fatty acids, including linoleic and α-linolenic acids, cannot be synthesised by mammals and must, therefore, be obtained from foods containing vegetable oils. For this reason, vegetable oils which are rich in mono- and/or polyunsaturated fatty acids, are generally regarded as having better nutritional qualities than those which are more enriched in saturated fatty acids. Vegetable oils are also actual or potential sources of a huge array of important industrial products, ranging from polymers, lubricants, paints and solvents, to cosmetics and pharmaceuticals (Murphy, 1996).

Finally, many plant secondary products play an important role, either in nutrition or in providing valuable products, e.g. for medicinal use. Examples include the glucosinolates, which are characteristic of the *Brassica* crops. These compounds are responsible for the flavour of mustard and the typical flavour and odour of *Brassica* vegetables. Glucosinolate derivatives have also been found to have important anticarcinogenic roles, both in epidemiological studies (Block *et al.*, 1992) and in laboratory-based feeding studies (Tawfiq *et al.*, 1995; Zhang *et al.*, 1994).

The biosynthesis and accumulation of all of these products, which determine the edible or industrial quality of a given crop, are largely or exclusively under genetic control. This allows us to manipulate such products in a particular crop plant in order to create an optimal profile for a particular end use. The application of genomics research is having on a particularly strong impact on our ability to manipulate the genetic traits responsible for crop quality.

GENOMICS AND PLANT RESEARCH

During the last 10 years, a great deal of plant research has been concentrated on a single model species, namely the cruciferous weed *Arabidopsis thaliana*. *Arabidopsis* has the virtue of containing a relatively small genome of only 100 Mb. This genome is arranged on 5 chromosomes, with relatively little repetitive DNA. In contrast, the model monocotyledon genome, i.e. that of rice, has a size of 500 Mb. The major crop species maize and barley have genome sizes of 2,500 and 5,000 Mb, respectively, while the hexaploid species, wheat, is 16,000 Mb. The relatively small genome size of *Arabidopsis* has made it the first plant target for a multinational genome sequencing project, which should cover the entire genome by 2003. Already, physical maps of the *Arabidopsis* chromosomes are becoming available (Schmidt, 1995) and the vast majority of the estimated 25,000 *Arabidopsis* expressed genes have already been identified as ESTs (expressed sequence tags) (Somerville, 1996).

In parallel with these genomic studies, saturating T-DNA and transposon tagged mutant libraries have been, or are being created, in addition to large numbers of chemically mutagenised lines of *Arabidopsis*. Ordered contiguous segments (contigs) of yeast artificial chromosomes (YACs) corresponding to much of the *Arabidopsis* genome have been created. This, in conjunction with detailed molecular maps based on RFLP, cDNA and other markers, now allows for positional cloning of genes which have been mapped to within an accuracy of about 1cM.

The coming challenge will be to utilise this formidable genetic resource based on a single, relatively simple, model plant, to effect improvements in the major dicotyledonous and monocotyledonous crop species. During the past few years, progress has been accelerating rapidly in transferring technologies and knowledge developed in *Arabidopsis* to some of the major crop species. Another encouraging development has been the recent description of the extensive synteny between all of the cereal genomes, which has revealed that they are composed of very similar chromosome segments. By rearranging these segments slightly and amplifying some of the repetitive DNA sequences contained within, it is possible to reconstitute the 56 different chromosomes found in wheat, rice, maize, sorghum, millet and sugar cane (Moore *et al.*, 1993; Moore, 1995). This means that a quantitative trait locus (QTL) which is mapped in a major cereal crops can also be localised by comparative genome analysis in all of the other major cereal crops including rice. Rice, with its relatively small genome size, is the *"Arabidopsis"* of the cereals world and efforts are currently underway to produce and map an EST library covering the entire rice genome.

FROM *ARABIDOPSIS* TO *BRASSICAS*

The *Brassica* spp and their close relatives include a number of important vegetable and oilseed species including broccoli, cabbage, mustard, radish, cauliflower and rapeseed. It has long been known that the *Brassica* spp can either be "diploid" or "amphidiploids" (U, 1935) as shown in Figure 1. More recently, evidence from several independent genomic studies, based largely on the analysis of molecular markers, has led to the conclusion that the "diploid" *Brassica* spp are probably really descended from hexaploid ancestors, i.e. they contain three largely intact but rearranged genomes, all of which are similar enough to have arisen from a single ancestor (Lagercrantz *et al.*, 1996; Scheffler *et al.*, 1997). Interestingly, the size of the putative ancestral diploid genome and its gene order is remarkably similar to the extant genome of *Arabidopsis thaliana* (Parkin *et al.*, 1997). It appears likely that an ancestral member of the Brassicaceae contained a relatively small genome, similar in size and gene order to that of present day *Arabidopsis* species. This ancestral diploid species may have undergone endo-duplication of its genome, or may have diverged to produce a number of related diploid species, which then hybridised to produce hexaploids, and then gave rise to the present "diploid" *Brassica* spp. Genome duplication is a common strategy which is associated with the ability of organisms to increase in both size and biochemical and therefore physiological complexity. Indeed, it has recently been reported that the yeast *Saccharomyces cerevisiae* probably underwent a genome duplication some 100 million years ago (Wolfe & Shields, 1997). It is proposed that this genome duplication may have been instrumental in the evolutionary adaptation of *Saccharomyces* to anaerobic fermentation.

It is interesting that the estimated date of the genome duplication corresponds to the time when Angiosperms (and their carbohydrate-rich fruits) became dominant components of the terrestrial flora.

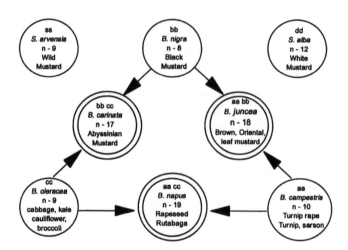

FIGURE 1: Genome and chromosome relationship of some important *Brassica* and *Sinapis* species (after U, 1935).

The *Arabidopsis* and *Brassica* genomes are probably related by a common ancestry as reflected in their syntenic similarities at the level of large chromosome segments. Recent work indicates an additional conservation of microsynteny, down to the level of individual genes and intergenic regions on a scale of 1-10 kbp (Sadanandom & Murphy, unpublished work). Analysis of known gene sequences indicates that homologous *Arabidopsis* and *Brassica* genes encode proteins which are about 85% identical in sequence. Even regulatory elements such as gene promoters are often highly conserved between *Arabidopsis* and *Brassica* spp (Sadanandom *et al.*, 1996). In our own laboratories and elsewhere, *Arabidopsis* promoters have been used to drive tissue-specific expression in transgenic *Brassica* plants and *vice versa* (Parmenter *et al.*, 1995).

The close similarity between these genomes means that molecular probes such as RFLPs and cDNAs can also be used interchangeably between these various *Arabidopsis* and *Brassica* species. A systematic comparison of the location of molecular markers in *Arabidopsis* and *Brassica* spp has already allowed for the mapping of chromosome segments covering most of the *Arabidopsis* genome to their three equivalent positions in the "diploid" *Brassica* genomes. This knowledge now allows us to move interchangeably between *Brassica* crops and *Arabidopsis* in order to identify, isolate and manipulate agronomically relevant genes, as discussed in the next section.

FROM QTL TO CLONED GENES

Many characters of agricultural importance in crop plants appear to be regulated by a large number of genes and do not, therefore, segregate into simple Mendelian ratios, as expected if only one or two genes were involved. Examples of such complex traits include canopy architecture, oil and protein yield in the seed, glucosinolate yield, flowering time and anther anthocyanin accumulation in maize caryopses. During the past 5 years, QTL mapping has proved to be a powerful tool for studying the inheritance of such complex characters. The method relies on genetic-linkage analysis and the availability of marker loci at a density of at least 20 cM throughout the genome. Molecular markers such as RFLPs have proved to be suitable for this purpose. QTL mapping allows for the estimation of the number of loci controlling a trait, their chromosomal location and the proportion of the phenotypic variation for which each locus is responsible. Probably the most surprising outcome of QTL mapping studies has been the realisation that even purely quantitative traits such as seed weight and height can often have 30%-50% of their phenotypic variation explained by a single locus (Doebley, 1993).

Examples of QTL analysis relating to the *Brassica/Arabidopsis* system include flowering time and glucosinolate yield. In both cases, a large proportion of the phenotypic variation in these characters was mapped to a single QTL in *Brassica*. Knowledge of the equivalent location within the *Arabidopsis* genome allowed for the rapid identification of *Arabidopsis* mutants with lesions in the relevant genes. Positional cloning can then enable the isolation of the candidate genes. In the case of the gene regulating flowering time, the encoded protein has a zinc-finger domain, characteristic of transcription factors (Putterill *et al.*, 1995; Coupland, 1997). Further analysis of transgenic plants in which the *CO* gene is expressed under the control of a strong viral promoter, indicates that transcriptional regulation of this gene is indeed an important determinant of flowering-time in response to photoperiod (Coupland, 1997).

FIGURE 2: From QTL mapping to new crop varieties (see next page)

Much of the variation in many apparently complex agronomic characters can often be explained by a small number of major quantitative trait loci (QTL). Such QTL can be localised to within 1-10 cM using molecular markers in segregating populations. In *Brassica* spp, the highly conserved genome synteny with *Arabidopsis* allows for the identification and fine mapping of the equivalent position of each QTL on the *Arabidopsis* genome. Availability of YAC (yeast artificial chromosome) and BAC (bacterial artificial chromosome) contigs and cosmid clones can then enable positional cloning of the gene corresponding to each QTL. The isolated gene can then be used to manipulate the original character in *Brassica* crop species. This knowledge also assists in the introgression, via marker-assisted selection, of such characters from wild relatives or other sources of useful genetic variation.

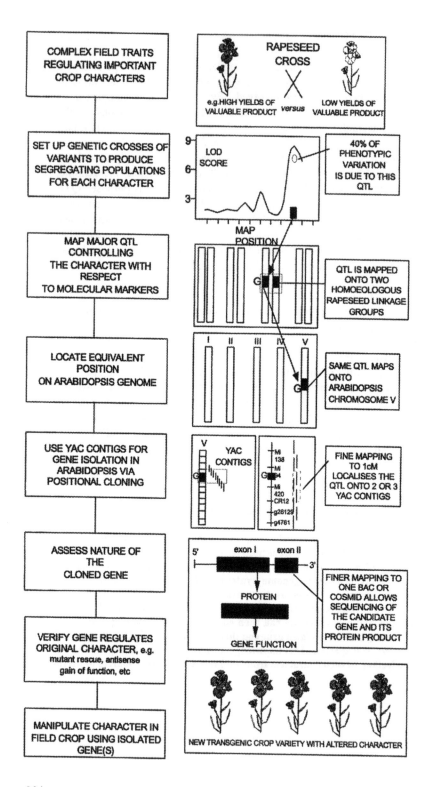

COMPLEX FIELD TRAITS REGULATING IMPORTANT CROP CHARACTERS

RAPESEED CROSS

e.g.HIGH YIELDS OF VALUABLE PRODUCT versus LOW YIELDS OF VALUABLE PRODUCT

SET UP GENETIC CROSSES OF VARIANTS TO PRODUCE SEGREGATING POPULATIONS FOR EACH CHARACTER

LOD SCORE

40% OF PHENOTYPIC VARIATION IS DUE TO THIS QTL

MAP POSITION

MAP MAJOR QTL CONTROLLING THE CHARACTER WITH RESPECT TO MOLECULAR MARKERS

QTL IS MAPPED ONTO TWO HOMOEOLOGOUS RAPESEED LINKAGE GROUPS

LOCATE EQUIVALENT POSITION ON ARABIDOPSIS GENOME

SAME QTL MAPS ONTO ARABIDOPSIS CHROMOSOME V

USE YAC CONTIGS FOR GENE ISOLATION IN ARABIDOPSIS VIA POSITIONAL CLONING

YAC CONTIGS

FINE MAPPING TO 1cM LOCALISES THE QTL ONTO 2 OR 3 YAC CONTIGS

ASSESS NATURE OF THE CLONED GENE

5' exon I exon II 3'

PROTEIN

FINER MAPPING TO ONE BAC OR COSMID ALLOWS SEQUENCING OF THE CANDIDATE GENE AND ITS PROTEIN PRODUCT

GENE FUNCTION

VERIFY GENE REGULATES ORIGINAL CHARACTER, e.g. mutant rescue, antisense gain of function, etc

MANIPULATE CHARACTER IN FIELD CROP USING ISOLATED GENE(S)

NEW TRANSGENIC CROP VARIETY WITH ALTERED CHARACTER

In both of the above cases, the progression from the identification and mapping of the major QTLs towards the isolation of the relevant gene from *Arabidopsis,* was considerably assisted by the availability of *Arabidopsis* lines containing single mutations in the gene of interest. Nevertheless, the availability of such mutations is not a prerequisite for this strategy. Fine mapping of QTLs down to less than 1 cM can allow for the positional cloning of relatively small chromosome segments corresponding to only 1-3 YAC contigs. The target gene can be localised still further by using BAC libraries, which contain an average insert size of only 100 kb. Several BAC libraries are available from *Arabidopsis,* containing in the region of 4-7 haploid genome equivalents. Even finer mapping, down to single cosmid clones is also possible. For example, filters are available, with duplicate samples of 37,000 cosmid clones from *Arabidopsis,* arranged in gridded arrays, with an average insert size of 17 kb. This library represents 7 haploid genome equivalents.

A number of binary vectors are now available for the insertion of either BACs or cosmids containing the putative target gene into plants, via techniques such as *Agrobacterium*-mediated transfer, vacuum infiltration, or microprojectile bombardment. The resulting transgenic plants are then screened for alterations in the phenotype of interest. In addition, the cosmid or BAC inserts are sufficiently small to allow for their sequencing via rapid throughput automated sequencers. In the future, even this latter step will be unnecessary once the sequence of the entire *Arabidopsis* genome is available. Following sequencing, potential open-reading frames are screened and candidate genes can be identified and used individually for transformation experiments, as described above and also in Figure 2.

Overlapping YAC contigs have also been produced, which cover about 50% of the rice genome, i.e. virtually all of that part of the genome which is not made up of repetitive DNA sequences. Further attempts are now underway to produce BAC contigs of the rice genome which will aid cloning efforts still further. Ideally, it would be useful to extend knowledge gained from the relatively facile *Arabidopsis* molecular genetics system directly to rice, and thereby to the other major cereal crops. Various laboratories are assessing the relatedness of the rice and *Arabidopsis* genomes, although the different codon usage between monocotyledons and dicotyledons will make it very difficult to use molecular markers or clone genes interchangeably between the two systems, as is possible between the *Brassica* spp and *Arabidopsis.*

DO THE MAJOR QTL ALWAYS ENCODE TRANSCRIPTION FACTORS?

The genomic approach to understanding and manipulating important plant characters has the particular virtue that it is not necessary to know in advance either the identity or even the possible function of the candidate gene. In a number of cases where this approach has been taken, the candidate gene was not part of a biosynthetic pathway directly involved in the character of interest. For example, the major gene *vp1,* which regulates anthocyanin biosynthesis in maize caryopses, encodes not an enzyme but a transcription factor, which interacts with the upstream promoter elements of some of the enzymes involved in the biosynthetic pathway. As mentioned above, the *CO* gene which regulates flowering time in *Arabidopsis* also appears to encode a transcription factor. Even more dramatically, it has recently been reported that a single gene, *teosinte branched 1* (*tb*1), was largely responsible for the alteration in plant architecture from the highly branched wild species *teosinte* and its relatively unbranched modern relative, maize (Doebley, 1997). The recently isolated *tb*1 gene encodes a protein with a putative nuclear localisation signal which suggests that it may play a role in transcriptional regulation.

Despite these and other examples however, it must not be concluded that all quantitative trait loci encode transcription factors, as has sometimes been suggested. For example, preliminary results from maize-breeding studies, indicate that an important QTL which regulates oil yield segregates with, and is probably located close to, the gene encoding acetyl-CoA carboxylase, which is the first committed step of fatty acid biosynthesis and hence of oil formation. Therefore, the use of genomics not only allows us to isolate genes regulating important plant developmental characters, but is also contributing greatly to our understanding of the biochemical and genetic regulation of these characters at the molecular level. A picture is gradually emerging whereby a small number of key genes can cause dramatic alterations in the phenotype of a given plant. So far, such genes have been found to encode either transcription factors, which can regulate entire metabolic pathways (eg anthocyanins) or developmental processes (e.g. flowering), or key enzymes that are often located at branch points of a metabolic pathway. It is evident that manipulation of such genes can effect dramatic phenotypic changes in the crop plant of interest. Witness the effect of the *tb*1 gene on the architecture of maize, as described above (Doebley, 1997), or the *cauliflower* gene in *Arabidopsis* which, as its name suggests, converts the plants into something resembling a miniature cauliflower (Kempin *et al.*, 1995). This is encouraging news for geneticists wishing to create completely novel phenotypes in existing crops or, indeed, to domesticate new crops from wild species.

FROM GENES TO CULTIVARS

Amongst its other virtues as a relatively simple model system for molecular genetics, *Arabidopsis* also has the distinct advantage of being relatively easy to transform. Amongst the major crop plants, relatively routine transformation methods have been established for several species such as rapeseed, potato and tomato, although once again, the efficiency of transformation, which normally involves the use of an *Agrobacterium* vector, varies considerably between different genotypes. The availability of efficient transformation systems for both *Brassica* spp and *Arabidopsis* means that the oilseed and vegetable *Brassica* crops are likely to be the first to benefit commercially from the application of genomic technology via *Arabidopsis*. Transformation systems have been developed for some of the major legume crops, including soybean and pea, but these remain relatively tedious, expensive, and restricted to a few of the larger research groups or companies. Cereal transformation has also made impressive strides during the past few years, although it still relies on specialised technologies, such as biolistics, and difficulties with regeneration of plantlets mean that it is also a relatively time-consuming and expensive process. Nevertheless, these technologies are rapidly developing and are even being applied to important perennial crop species such as eucalypts and oil palm.

During the next decade, relatively efficient transformation methods will probably be available for all of the major crop species. This will allow for the introduction of genes controlling important developmental traits, such as protein or oil quality, once these are identified via QTL mapping and positional cloning, as outlined above. One obvious potential problem is that QTLs which regulate a character in *Brassica* spp and *Arabidopsis* may not necessarily regulate the same character in a very distantly related species, such as wheat or rice. However, there are now indications that many important developmental processes, such as flowering, meristem development and even quantitative traits like height, are probably regulated by very similar genes, even in very distantly related plant species. Indeed, some of the homeotic genes regulating pattern formation and other developmental processes in plants, encode transcription factors that are very similar to those regulating pattern formation and cell fate in species as diverse as *Drosophila*, yeast and *Homo sapiens*.

MANIPULATION OF QUALITY TRAITS

Agriculturally important quality traits, such as seed oil and protein composition have been amongst the first to be manipulated by genetic engineering. One of the major reasons for this was a common perception that such traits would be determined by a very small number of genes, i.e. the insertion of one or two transgenes into a crop would be likely to result in a dramatic change in the quality phenotype of interest. This meant that such manipulations should be technically relatively straightforward, achievable within a few years, and not prohibitively expensive. However, the findings from the first wave of transgenic rapeseed varieties, demonstrate that some of the characters determining fatty acid quality in seeds behave as quantitative traits, often involving at least 3 or 4 major genes (Voelker *et al.*, 1996; Murphy, 1996). Experience with altering lauric acid, petroselinic acid or ricinoleic acid phenotypes indicates that, while 20-40% levels of a given novel fatty acid in the seed oil may be possible following the insertion of a single transgene, the achievement of more commercially desirable levels of 60-90% requires the presence of 3 or 4 different transgenes. To complicate matters still further, the presence of "unusual" fatty acids, such as lauric (Ohlrogge & Eccleston, 1996) or petroselinic (Bowra, Fairbairn & Murphy, unpublished results), has been found to stimulate lipid breakdown via the β-oxidation and glyoxylate cycle pathways, leading to a wasteful futile cycling. Clearly, there is a need for more research into the genetics and biochemistry of lipid metabolism in seeds so that more targeted and rational strategies can be devised for oil quality manipulation using transgenic methods.

Despite these concerns, it should be possible to produce a seed oil with commercially useful levels of virtually any desired fatty acid, or a seed protein with the desired balance of essential amino acids within the next decade or so. Important advances in understanding the regulation of starch granule formation, also make it possible to envisage the targeted manipulation of starch quality in major crop species. Other quality traits involving secondary metabolic pathways, e.g. glucosinolates, are also being targeted for transgenic manipulation.

Another important aspect of crop quality is the composition of the harvested material. For example, in seeds, the major products are either starches, proteins or oils. Different crop types tend to accumulate different ratios of these storage products, e.g. cereals tend to accumulate a lot of starch, while oilseeds tend to accumulate oils. The ratios of starch:protein:oil are largely determined genetically and can, therefore, be manipulated beyond the natural variation of the particular species by using transgenic methods. For example, most of the major oilseed crops contain about 20-45% $^W/_W$ oil in their seeds. Despite decades of efforts by breeders, it has not been possible to obtain significantly higher levels of seed oil. However, there are many other oil-rich plants which contain as much as 60-76% $^W/_W$ oil in their seeds (Murphy, 1996). This implies that there is no fundamental biological reason why yields of certain seed oils cannot be increased to almost double the present levels. Efforts are now underway to identify and manipulate QTLs regulating oil yield in crop plants of agronomic interest. The same strategy can be used for the manipulation of starch or protein yield in seeds.

COMMERCIAL PERSPECTIVE

The enormous advances made in plant genomic research during recent years have now led to an increased participation by the commercial sector. Plant genome research and development has always been strongly supported by large companies with an interest in agribusiness - particularly those involved in plant protection, nutrition and seed production. Further stimulation and cross-fertilisation has come from the presence of numerous small biotechnology companies, many of which originated as spin-offs from laboratories in universities or research institutes. The research has also benefited from a generally enlightened public sector funding policy, which has allowed for the implementation of several large transnational programmes, some of the best examples of which are the *Arabidopsis* EST and genome sequencing projects. A very important principle amongst *Arabidopsis* researchers has been the rapid dissemination of information and universal availability of resources, such as EST clones and YAC contigs, throughout the scientific community, whether in the public or private sector. This has benefited both academic researchers and commercial organisations. More recently, however, there has been a consolidation of several small-to-medium sized biotechnology companies following their acquisition by some of the larger agribusiness concerns. Examples include the acquisition of Calgene by Monsanto, at a valuation of about $470 million, the acquisition of Mogen by Zeneca for $75 million, the acquisition of 75% of PGS by AgrEvo for $550 million, and the acquisition of a 20% interest in Pioneer by DuPont for $1.7 billion. In all of the cases, the biotechnology companies held important proprietary technologies relating to the manipulation of quality in transgenic crops. These developments need not inhibit future R&D, providing that suitable cross-licensing arrangements can be made for the use of these proprietary technologies, and that the larger agribusiness concerns maintain the kind of open dialogue with the rest of the research community that was part of the culture of many of the smaller biotechnology companies.

FUTURE PERSPECTIVES

It has been stated that we are now living in the "golden age" of biology and this public perception has much to do with the enormous impact of genomic research on medicine and agriculture. Genomics is providing us with powerful tools to understand and manipulate many of the basic developmental processes in crop plants. It also allows us to intervene at the level of single genes, in order to effect the precise manipulation of a single compound, e.g. a valuable fatty acid in the seed oil. Nevertheless, genomics alone is insufficient for anything other than an empirical approach to manipulation of, for example, quality traits. It is essential that the genetical strategy is complemented by physiological and biochemical studies. This will require a robust collaboration between the public and private sectors. Large companies like Monsanto, Dupont and Pioneer are investing heavily in genomics R&D and several multinational public sector projects, e.g. *Arabidopsis* and rice sequencing, are underway. These activities are drawing large amounts of funding (and therefore people and expertise) into applied genomics. In the coming decade it is important that we retain the capacity to exploit the plethora of information that will be generated from genomic research. We should continue to develop appropriate informatics systems to distill the sequence, mapping and structural data into biologically useful forms. Perhaps even more important will be to link genomic information with biological functions at all the relevant hierarchical levels from enzyme and pathway to organism and population. These are impressive challenges but the potential rewards, particularly for agriculture, are enormous.

REFERENCES

Batchelor, S., Booth, E., Entwhistle, G., Walker, K., ap Rees, T., Hacking, A., Mackay, G. & Morrison, I. (1996) Industrial markets for UK-grown crop polysaccharides. *Research Review No. 32*. London: Home Grown Cereals Association.

Block, G., Patterson, B. & Suber, A. (1992) Fruit, vegetables, and cancer prevention - a review of the epidemiologic evidence. *Nutrition and Cancer* **18**, 1-19.

Coupland, G. (1997) Regulation of flowering by photoperiod in *Arabidopsis*. *Plant Cell and Environment* **20**, 785-789.

Doebley, J., Stec, A. & Hubbard, L. (1997) The evolution of apical dominance in maize. *Nature* **386**, 485-488.

Doebley, J. (1993) Genetics, development and plant evolution. *Current Opinions in Genetic Development* **3**, 865-

Kempin, S.A., Savidge, B. & Yanosky, M.F. (1995) Molecular basis of the cauliflower phenotype in *Arabidopsis*. *Science* **267**, 522-525.

Lagercrantz, U., Putterill, J., Coupland, G. & Lydiate, D. J. (1996) Comparative mapping in *Arabidopsis* and *Brassica*, fine- scale genome collinearity and congruence of genes controlling flowering time. *Plant Journal* **9**, 13-20.

Moore, G., Devos, K.M., Wang, Z. & Gale, M.D. (1995) Grasses, line up and form a circle.*Current Biology* **5**, 737-739.

Moore, G. (1995) Cereal genome evolution: pastoral pursuits with "Lego" genomes. *Current Opinions in Genetics and Development* **5**, 717-724.

Murphy, D.J. (1996) Biotechnological improvement of oil crops. *Trends in Biotechnology* **14**, 206-213.

Ohlrogge, J.B. & Eccleston, V.S. (1996) Coordinate induction of pathways for both fatty acid biosynthesis and fatty acid oxidation in *Brassica napus* seeds expressing lauroyl-ACP thioesterase. In: *Proceedings of the 12th International Symposium on Plant Lipids, Toronto.* Abstract V2.

Parkin, I.A., Cavell, J., Oldknow, J., Lydiate, D.J. & Trick, M. (1997) Genome synteny between *Arabidopsis thaliana* and *Brassica napus*. In: *John Innes Centre Annual Report*, pp13-14.

Parmenter, D.L., Boothe, J.G., van Rooijen, G.J.H., Yeung, E.C. & Moloney, M.M. (1995) Production of biologically active hirudin in plant seeds using oleosin partitioning. *Plant Molecular Biology* **29** 1167-1180.

Putterill, J., Robson, F., Lee, K. & Simon, R. (1995) The *CONSTANS* gene of *Arabidopsis* promotes flowering and encodes a protein showing similarities to zinc-finger transcription factors. *Cell* **80** 847-858.

Sadanandom, A., Piffanelli, P., Knott, T., Robinson, C., Sharpe, A., Lydiate, D.J., Murphy, D.J. & Fairbairn, D. (1996) Identification of a peptide methionine sulfoxide reductase gene in an oleosin promoter from *Brassica napus*. *Plant Journal* **10**, 235-242.

Scheffler, J., Sharpe, A.G., Schmidt, H., Sperling, P., Parkin, I.A.P., Luhs, W., Lydiate, D.J. & Heinz, E. (1997) Desaturase multigene families of *Brassica napus* arose through genome duplication. *Theoretical and Applied Genetics* **94**, 583-591.

Schmidt, R., West, J., Love, K., Lenehan, Z., Lister, C., Thompson, H., Bouchez, D. & Dean, C. (1995) Physical map and organization of *Arabidopsis thaliana*. *Science* **270**, 480-483.

Smith, A. & Martin, C. (1993) Starch biosynthesis and the potential for its manipulation. In: *Biosynthesis and Manipulation of Plant Products*, D. Grierson (Ed) London: Blackie Academic & Professional, Chapman and Hall, pp1-54.

Somerville, C. (1996) The physical map of *Arabidopsis* chromosomes. *Trends in Plant Science* **1**, 2.

Tawfiq, N., Heaney, R., Plumb, J., Fenwick, G., Musk, S. & Williamson, G. (1995) Dietary glucosinolates as blocking agents against carinogenesis - glucosinolate breakdown products assessed by induction of quinone reductase activity in murine HEPA1C1C7 cells. *Carcinogenesis* **16**, 1191-1194.

U, N. (1935) Genomic analysis in *Brassica* with special reference to the experimental formation of *Brassica napus* and peculiar mode of fertilisation. *Japanese Journal of Botany* **7**, 389-452.

Voelker, T.A.,Hayes, T.R., Cranmer, A.M., Turner, J.C. & Davies, H.M. (1996) Genetic engineering of a quantitative trait - metabolic and genetic parameters influencing the accumulation of laurate in rapeseed. *Plant Journal* **9**, 229-241.

Wolfe, K.H. & Shields, D.C. (1997) Molecular evidence for an ancient duplication of the entire yeast genome. *Nature* **387**, 708-713.

Zhang, Y., Kensler, T.W., Cho, C.G., Posner, G.H. & Talalay P. (1994) Anticarcinogenic activities of sulforaphane and structurally related synthetic norbornyl isothiocyanates. *Proceedings of the National Academy of Science, USA* **91**, 3147-3150.

POSTERS

OUTLINE OF THE U.S. NATIONAL CORN GENOME INITIATIVE.

J.S. McLAREN

Inverizon International Inc., Chesterfield, MO 63017, USA

R. MUSTELL

National Corn Growers Association, St. Louis, MO 63141, USA

ABSTRACT

The National Corn Genome Initiative (NCGI) is supported by a consortium of private companies and public sector organizations, and has an objective of generating a corn (maize) genome map to provide a resource that will support and stimulate sustainable economic advances in crop production.

The past and the future have already been written, in the genetic code....
If only we could afford to read the book.

INTRODUCTION

World population continues to expand and improvements in living standards in the developing countries have resulted in an exponential increase in the demand for animal protein. Crop breeding has allowed steady advances in production, and some modification of desired quality attributes. For example, in corn, yield has increased at about 101 pounds/acre/year (1.4%/acre/year) since the adoption of hybrids, and genetically-selected high oil corn can provide enhanced feed energy levels without the need to add animal fat. However, in order to meet the global demands of the 21st century we need a breakthrough in the rate of crop improvement and utilization, while being respectful of the growing environmental concerns over intensive production systems. Because of the importance of corn in the global feed and food chain, the NCGI believes that creating a corn genome map would have significant positive consequences in meeting these challenges.

THE GOAL

The goal is to implement a directed and coordinated program to clone, sequence and map the approximately 50-80,000 genes which control growth, development, yield and quality in corn (*Zea mays*). There would also be associated work related to gene expression information, maintaining physical stocks, and the development of a computer-based informatics system to store, retrieve and utilize data. In order to accomplish this goal, the NCGI has requested that the U.S. Federal Government appropriate $143 million over a five year time-frame.

While genetic maps exist, marker-based maps are becoming more common, and some sequences have been determined, a coordinated effort to map the whole corn genome does not exist today. It is envisaged that the basic genomic information obtained from a corn map will be compiled in a central database, for which a foundation exists in a present USDA program, and that such information will be accessible for application to corn and other world crops.

RELATED WEB SITES

Additional details may be found at:
NCGI URL is http://www.inverizon.com/ncgi/
MaizeDB URL is http://www.agron.missouri.edu/

COMPARATIVE GENE MAPPING WITH THE CHICKEN - CLUES TO OUR VERTEBRATE ANCESTRAL GENOME

J. SMITH, C.T. JONES, I.R. PATON, A.S. LAW AND D.W. BURT

Department of Molecular Biology, Roslin Institute, Roslin, Midlothian, EH25 9PS, UK

ABSTRACT

The ability to locate trait genes in poultry is dependent upon the generation of genetic linkage maps. To identify such genes, a comparative positional candidate gene approach can be undertaken, the success of which is thus highly dependent on the construction of comparative maps.

INTRODUCTION

Genes associated with a particular QTL (quantitative trait locus) can be isolated by identifying candidate genes within a region of interest by comparative mapping techniques. The location of a gene can be predicted by the conservation of synteny and gene order between species. This conservation of genomic regions can also help us trace the evolutionary pathways taken by different species, and to examine in more detail how different genomes have diverged.

DISCUSSION

For genes in different species to be considered homologous, they should have a similar nucleotide sequence, be able to show cross-hybridization to the same molecular probe and also have a conserved map position. The presence of homologous genes is evidence of homology between genomes of two separate species, although the existence of conserved synteny and gene order is stronger proof. Conserved gene order is the syntenic association of three or more contiguous (not interrupted by different chromosome segments) homologous genes, in the same order in two separate species. For a review of chicken genome mapping, see Burt *et al.*, 1995.

An example of how conserved map position has been used to define homologous genes is shown in figure 1. There are two MIF (macrophage inhibition factor) genes mapped in the chicken. By comparative mapping analyses, it is seen that the gene mapping to the chicken linkage group E18C15 is the true homologue of the human and mouse genes.

Conservation of autosomal segments of the chicken, human and mouse was analysed and the number of chromosomal rearrangements which have taken place in these species since the divergence of birds and man has been calculated (Nadeau and Taylor, 1984). Comparison of homologous segments revealed 74 between chicken and human, 217 between chicken and mouse and 170 between human and mouse. It is a great surprise to find that there are estimated to be only 50 rearrangements between chicken and human, whilst there are around 190 between chicken and mouse and 150 between human and mouse. How the genomes have therefore diverged is shown in figure 2.

ACKNOWLEDGEMENTS

This work was supported by the Biotechnology and Biological Research Council, UK; the Ministry of Agriculture, Fisheries and Food, UK and by EC grant no. BIO4-CT95-0287, as part of the Chickmap project.

	Chicken		Mouse		Human	
CRYBB1	E18C15	15.88	5	59.50	22	q11.2
CRYBA4	E18C15	15.88	5	59.00	22	q11.2-q13.1
IGLV	E18C15	19.89	16	13.00	22	q11.2
IGLL	E18C15	19.89	16	13.00	22	q11.2
MIF	E18C15	19.89	10	42.25	22	q11.2
MIFL1	2	41.04	?		?	

FIGURE 1: MIF on the chicken linkage group E18C15 is homologous to the gene on human chromosome 22.

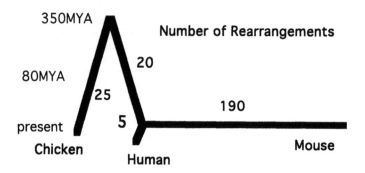

FIGURE 2: The divergence of the chicken, human and mouse genomes. (MYA - millions of years ago).

REFERENCES

Burt, D.W., Bumstead, N., Bitgood, J.J., Ponce de Leon, F.A. and Crittenden, L.B. (1995) Chicken genome mapping: a new era in avian genetics. *Trends in Genetics*, **11**, 190-194.

Nadeau, J.H. and Taylor, B.A. (1984) Lengths of chromosomal segments conserved since divergence of man and mouse. *Proceedings of the National Academy of Sciences (USA)*, **81**, 814-818.

PRIMER EXTENSION FROM TWO-DIMENSIONAL OLIGONUCLEOTIDES GRIDS FOR DNA SEQUENCE ANALYSIS

A. METSPALU[1,2], H. SAULEP[1], A. KURG[1], N. TÕNISSON[1]

[1]*Tartu University, Institute of Molecular and Cell Biology, Estonian Biocentre, 23 Riia St., EE2400 Tartu, Estonia;* [2]*Molecular Diagnostics Centre at Tartu University Children's Hospital, 3 Oru St., Ee2400 Tartu, Estonia*

J.M. SHUMAKER

Baylor College of Medicine, Department of Molecular and Human Genetics ,One Baylor Plaza, Houston, Texas 77030, USA

ABSTRACT

A method is described for sequence analysis of nucleic acids in which arrayed primer extension (APEX) was used. A two-dimensional oligonucleotide array (DNA chip) was fabricated on a glass support. Oligonucleotides (ONs) were used to scan, in single base increments, the DNA sequence of interest. ONs were immobilized via a 5' amino link to epoxyactivated glass surface. Disease gene exons were amplified by PCR. Single-stranded DNA served as a template for the primer extension reaction on the DNA chip using fluorescently labeled ddNTPs and T7 DNA polymarease. Each ON on the DNA chip could only be extended by one base. A short fragment of the p53 gene was resequenced in this study as the proof of the principle of this technology. Furthermore, APEX could be applied not only to resenquencing for mutation detection, but also to quantitative analysis of RNA expression.

INTRODUCTION

The Human Genome Project is projected to generate nucleotide sequences of all human genes early in next decade. The availability of gene sequence information will facilitate the identification of mutations that either cause genetic disorders or predisposition to multifactorial diseases. Clearly, more efficient DNA resequencing and analysis methods are needed to exploit fully this new set of DNA sequence information. A radically new concept of DNA sequencing was put forward about a decade ago by several groups (for review see Mirzabekov, 1994 and Southern, 1996). A technical breakthrough for the use of DNA chips was the introduction of photolithographic combinatorial *in situ* synthesis of ON arrays (Fodor *et al.*, 1991). However, the original goal for rapid *de novo* DNA sequencing has not been (and may never be) realized. Recent studies have either introduced new goals for oligonucleotide arrays (Schena *et al.*, 1995; Shumaker *et al.*, 1996; Pastinen *et al.*, 1997) and/or modified the goal of the original DNA chip concept (Fodor 1997).

METHODS

5'-Amino-modified 15-mer oligonucleotides, complementary to 37 bases in exon 7 of the p53 gene were synthesized by Pharmacia Biotech (Uppsala, Sweden) and spotted by hand together with controls onto the microscope glass slide as a 7x6 grid. Glass slides were epoxysilanized before use as described previously (Shumaker *et al.*, 1996). ONs were diluted before to 50 μM in 0.1 N NaOH and 0.5 ml was pipetted to each spot. As a template, synthetic 63-mer ON was used. Primer extension reaction was performed in 40 μl: [5 units KlenTaq DNA polymerase (Ab Peptides, St. Louis, USA) in PC2 buffer (supplied with the enzyme) 12.5 mM $MnCl_2$, 7.5 mM DTT, 0.5 mg/ml BSA 20 pmoles of ssDNA template (62-mer ON) and fluorescent ddNTPs 0.25 μM of each (DuPont/NEN)]. The reaction mix was pipetted onto the grid, covered by the cover slip and incubated at 60°C for 1 min. The grid was then cooled down to 42°C, washed briefly with 70°C water, covered with 10 ml of Slow Fade (Molecular Probes) under the clean coverslip and scanned on a FluorImager 575.

RESULTS AND DISCUSSION

The results of the APEX reaction with the p53 grid are shown in Fig. 1. Four chips were used, one for each labeled ddNTP: a "one color - four chips" format. Note that the DNA sequence can be read from left to right counting the dark spots at each position. The development of a four color detector provides for a "four color - one chip" format.

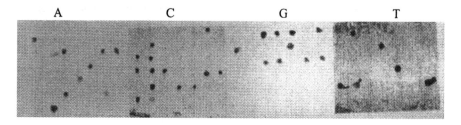

FIGURE 1. DNA chip resequencing exon 7 fragment of the human p53 gene. First top three lines can be read as follows: ATGGGCGGCATGAACCGGAGG

APEX requires only one ON to identify one base. In contrast, the Affymetrix approach requires at least four, and frequently as many as 14, oligonucleotides per identified base (Hacia *et al.*, 1996). Moreover, power of discrimination between genotypes was 9-27 times better in APEX, compared to the allele-specific oligonucleotide (ASO) method (Pastinen *et al.*, 1997). Inherent in the APEX approach is the property of the DNA polymerase to discriminate against 3'-end mismatches, therefore ensuring perfect Watson and Crick basepairing at the 3'-end of the oligo-template complex. Unfortunately, high density ON grids synthesized by the efficient photolithographic method (Fodor *et al.*, 1991) are not applicable for APEX assay, since oligonucleotides are attached to the glass support via 3'-end, thereby prohibiting the 5'-3' activity of template-dependent DNA polymerases

ACKNOWLEDGMENTS

This work was supported by the Estonian Science Foundation grant no. 2492 and EC grant no. CIPA-CT94-0148.

REFERENCES

Fodor, S.P.A. (1997) Massively parallel genomics. *Science* **277**, 393-395.

Fodor, S.P.A., Read, J.L., Pirrung, M.C., Stryev, L., Lee, A.T. and Solas, D. (1991) Light-directed, spatially addressed parallel chemical synthesis. *Science* **251**, 767-773.

Hacia, J.G., Brody, L.G., Chee, M.S., Fodor, S.P.A. and Collins, F.S. (1996) Detection of heterozygous mutations in BRCA1 using high density oligonucleotide arrays and two-colour fluorescence analysis. *Nature Genetics* **14**, 441-447.

Mirzabekov, A.D. (1994) DNA sequencing by hybridization - a megasequencing method and a diagnostic tool? *Trends in Biotechnology.* **12**, 27-32.

Pastinen, T., Kurg, A., Metspalu, A., Peltonen, L. and Syvänen, A.-C. (1997) Minisequencing: A specific tool for DNA analysis and diagnostics on oligonucleotide arrays. *Genome Research* **7** ,606-614.

Schena, M., Shalon, D., Davis, R.W. and Brown, P.O. (1995) Quantitative monitoring of gene expression patterns with a complementary DNA microarray. *Science* **270**, 467-470.

Shumaker, J.M., Metspalu, A. and Caskey, T.C. (1996) Mutation Detection by Solid Phase Primer Extension. *Human Mutation* **7**, 346-354.

Southern, E.M. (1996) DNA chips: analyzing sequence by hybridization to oligonucleotides on a large scale. *Trends in Genetics* **12**, 110-115.

ADDITIONS AND EXPANSIONS TO THE FLEXYS™ LABORATORY ROBOT SYSTEM

G. McKEOWN

PBA Technology Ltd, Unit 3, Forge Close, Eaton Socon, St Neots, Cambs, PE19 3TP, UK, gavin@pbatech.co.uk, http://www.pbatech.co.uk

ABSTRACT

The Flexys™ Robot was designed and developed in conjunction with The Sanger Centre, Hinxton, UK. The result of this collaboration is an adaptable system, suitable for many tasks in the field of molecular biology research. The robot is sold world-wide under licence by PBA Technology Ltd and their distributors. This paper will describe some of the developments which have recently taken place on the Flexys, designed to improve throughput and capacity.

INTRODUCTION

Previous automation in the molecular biology laboratory has been largely targeted at a single operation such as colony picking (Jones *et al.*, 1992) or gridding (McKeown *et al.*, 1994). These machines tend to be inflexible, quickly becoming obsolete as the methods used in the laboratory change. The Flexys Robot is designed around the core concept of modularity. This modularity extends right through the hardware, the electronics and the software, and enables the robot to be changed and developed over time, thus avoiding the risk of obsolescence. The modular design enables the user to change from automated colony and plaque picking to automated high density gridding in a matter of minutes.

THE FLEXYS

The Flexys robot is based around an aluminium frame of welded construction with precision slides for the main axes. Motive power is provided by five phase stepper motors directly driving toothed belts. A key feature of the modular design is the distributed control system (Stewart *et al.*, 1995). This places the control electronics for each active part of the robot close to the item being controlled, rather than in the host computer or a centralised electronics rack. Each distributed control board is part of a family of custom designs, each based around a PIC 17C42 microcontroller (Microchip Technology Inc., Chandler, Arizona). Each member of the family is designed to perform a particular task or tasks, but all share a common interface. This allows new functionality to be designed and added into the system with ease. The recent additions to the Flexys system are discussed below.

The AutoLoader

The AutoLoader extends the unattended run time of the standard Flexys. The user enters the run parameters in the usual way and, with guidance from the control system, loads all the plasticware required into the AutoLoader stack. The AutoLoader then exchanges carriers as and when required throughout the run with no further user intervention. The AutoLoader is already integrated with all existing Flexys software and its high capacity of 120 microplates will enable new, plate-intensive applications such as Re-arraying to be developed.

The 24-pin picking head

The 24-pin picking head boosts the picking rate of the Flexys Colony & Plaque Picker application to in excess of 1,500 picks per hour. Each of the 24 solenoid controlled needles is independently calibrated using an automatic procedure to ensure picking accuracy of over 99%. The 24-pin picking head can be used to pick colonies or plaques from any type of plasticware, and can inoculate standard microplates in 96- or 394-well densities or 96 density deep well boxes.

The Flexys 348 Wellfiller

In order to complement the high throughput picking capability of the Flexys and AutoLoader, PBA Technology have designed and developed the Flexys 384 Wellfiller. This stand-alone device is capable of dispensing media into the wells of microplates in either 96- or 384-well densities and also into 96 density deep well boxes using an 8-way manifold. Volumes from 10μl to 2ml are set through an easy-to-use keypad and liquid crystal display. The Wellfiller dispenses with very high accuracy, thanks to a swept volume pump design, and an RS232 port is included to enable the Wellfiller to be incorporated into larger automation systems.

REFERENCES

Stewart, M., Watson, A., McKeown, G., Karuntaratne, K. and Davies, R., (1995) A general purpose instrument control system and its application to a flexible laboratory robot. *Laboratory Robotics and Automation*, **Vol 7 (2)**, 85-91.

McKeown, G. (1994) A new flexible automation system. *Genome Digest*, **Vol 1 (5)**, 7.

McKeown, G., Watson, A., Karuntaratne, K. and Bentley, D. (1994) High throughput filter preparation robot. *Genome Science and Technology*, **1**, 56.

Jones, P., Watson, A., Davies, M. and Stubbings, S. (1992) Integration of image analysis and robotics into a fully automated colony picking and plate handling system. *Nucleic Acids Research*, **20**, 4599-4606.

Subject Index